INDUSTRIAL RESOURCE UTILIZATION AND PRODUCTIVITY

INDUSTRIAL RESOURCE UTILIZATION AND PRODUCTIVITY

Understanding the Linkages

Anil Mital, PhD
University of Cincinnati

Arunkumar Pennathur, PhD
University of Texas at El Paso

MOMENTUM PRESS

Momentum Press, LLC, New York

Industrial Resource Utilization and Productivity: Understanding the Linkages

First published in 2010 by
Momentum Press®, LLC
222 East 46th Street, New York, N.Y. 10017
www.momentumpress.net

ISBN-13: 978-1-60650-130-6 (hard back, case bound)
ISBN-10: 1-60650-130-5 (hard back, case bound)

ISBN-13: 978-1-60650-132-0 (e-book)
ISBN-10: 1-60650-132-1 (e-book)

DOI forthcoming

Cover Design by Jonathan Pennell
Interior Design by Scribe, Inc. (www.scribenet.com)

First Edition August 2010

10 9 8 7 6 5 4 3 2 1

Printed in Taiwan

DEDICATED TO

Mrs. Chetna Mital, MD, and Ms. Priya Pennathur, PhD,
and all those in the forefront of the fight against global warming.

CONTENTS

Part IV. Humans

Part V. Techniques

ACKNOWLEDGMENTS

First and foremost, we would like to thank Mr. Joel Stein, the publisher, for his tireless efforts and encouragement in getting our manuscript into shape for production. The chapters in the book are extended versions of the *First International Conference on Enhancing Resource Effectiveness in the Workplace*, which was held in Kathmandu, Nepal, in 2009. The conference was hosted by the *International Journal of Industrial Engineering—Theory, Applications, and Practice*. The editors thank Dr. Anand Subramanian, Dr. Bheem Kattel, Mr. R. Acharya, Mr. R. K. Sharma, Mr. Suraj Vaidya, and the Federation of Nepalese Chambers of Commerce and Industry for helping with the organization of this unique conference. The editors also thank all the authors who wrote extended versions of their book chapters with such a short notice.

Anil Mital, PhD
Arunkumar Pennathur, PhD

PREFACE

Our blue planet, the only life-sustaining planet we know of, is in peril due to global warming caused by greenhouse gases. And yet the 192 countries (various sources have quoted the number of participating countries from 192 to 196) that were involved with the United Nations (UN) meeting in December 2009 in Copenhagen, Denmark, while acknowledging the need to reduce the emission of greenhouse gases, failed to reach an agreement on how to deal with the global warming issue. The UN-sponsored conference only agreed to hold the increase in global temperature to below 2 °C by the year 2020. The developed countries also agreed to mobilize $100 billion a year by 2020 to address the needs of developing countries to curb greenhouse gases. No specific time-bound reduction in greenhouse gases was agreed upon. In the absence of such targets, life as we know it may not survive.

It is widely acknowledged that 83% of the greenhouse gases are caused by energy-generating activities that use fossil fuels such as coal and oil. Additional global warming is the result of machinery such as automobiles. Besides providing electricity to households, generated power also allows us to undertake the industrial activities that form the backbone of our modern society. Electric power, coupled with technology, provides for human needs. These needs, while not equal in all parts of the world, grow with economic development.

The provision for human needs requires the use of resources, renewable and nonrenewable as well as natural and man-made, for generating power, products, and services. Generally speaking, from an engineering viewpoint, resources are traditionally classified as human, materials, equipment, and capital. Recently, water and energy have been added to this list. Many of these resources, such as minerals and fossils, are becoming scarce as the economic development across the globe accelerates. This is particularly true for nonrenewable resources like coal and oil, which cannot be recycled. It now takes the movement of 4 tons of earth to produce a single barrel of oil, indicating a decline in the global production of oil. Other resources, such as humans and water, though available in abundance, have a different kind of problem associated with them. For instance, two-thirds of our planet surface is covered with water and yet large parts of the world are inflicted with drought conditions, and, thus, people do not have adequate supplies of clean water for drinking and irrigation. Likewise, Africa and Asia, though heavily populated, do not have the educational and training facilities to prepare their workforce for the production of goods and provision of services.

For a variety of reasons, some of which have been alluded to previously, we find ourselves at a juncture where we must decide to undertake measures to optimize the use of resources. That is, we must minimize the use of nonrecyclable natural resources, such as oil and coal; enhance the utilization of recyclable natural resources, such as metals; better educate and train our biggest resource, that is, humans; reduce the consumption of water in activities that represent inequitable distribution, such as the growing of crops in arid and semiarid areas that require intense irrigation; and increase the use of renewable sources of energy, such as solar and wind.

In order to accomplish all these tasks, a number of things must be accomplished. One requirement is to enhance the utilization of resources in the workplace. Keeping that in mind, the editors organized the *First International Conference on Enhancing Resource Effectiveness in the Workplace* in Kathmandu, Nepal, in August 2009. This book is the outcome of that conference and includes 14 chapters divided into 5 parts: energy, materials, equipment and maintenance, humans, and techniques. While it would have been desirable to have more chapters in each category and at least some chapters in additional parts, such as "water" and "capital," this effort is not considered to be comprehensive. Rather, we hope to draw the attention of the engineering community to the relevance of enhancing resources and the need of the hour.

In the first category, Asfour and Eltoukhy discuss the sustainability of manufacturing operations. Their contention is that sustainability can only be achieved through energy efficiency and careful product and packaging design that allow recycling, reuse, and rebuilding. As a stand-alone in this section, it provides general guidelines for sustainability. There are two chapters in the second section, which deals with materials. Following the sustainability theme, Desai and Mital present a methodology for disassembling products that encourages recycling and reuse. Enhancing durability, a critical cog in the sustainability wheel, is analyzed by Jerath and Moretti as they discuss ways to prevent steel in bridges from corroding quickly.

The third section of the book includes two chapters, which deal with enhancing durability through maintenance. While Desai and Mital propose a procedure that allows the evaluation of equipment design for enhanced maintainability, Jha, Devkota, and Kattel utilize an optimization approach to maximize service quality.

Chapters dealing with the human resource are grouped in the fourth section. The interdependency between humans and technology is explored by Mital, Pennathur, and Senapati. They infer that the weaknesses of the one are complemented by the strengths of the other and the interdependency is essential to get the most out of both resources. Managing the human resource and designing appropriate work systems are tackled by Ikeda and Masuzawa in the following chapter. This is followed by the development of an allocation model by Gilad and Khabia for hospital nurses, a resource in short supply worldwide.

The last section of the book groups six chapters, which discuss a wide variety of techniques—from developing an improved procedure for modeling industrial processes by

García and Noriega to minimizing waste of natural resources by Valles-Chavez and Sanchez. Specific procedures for improving demand forecasting, managing shelf space (a rare commodity in retail business), managing maintenance information, and modeling the allocation of resources for improving productivity are dealt with in the other chapters in this section.

We realize that enhancing the utilization of resources in the workplace is a huge topic for the variety and nature of the workplace, and the types of resources one must deal with present a big challenge. Nevertheless, the current need of the planet and its inhabitants is dire, and one must begin somewhere. This book represents the beginning of our efforts, not the end. We hope to follow up this effort in the future by educating the engineering community on how to enhance the utilization of resources in the workplace. In this endeavor, we are grateful to our publisher, Mr. Joel Stein, and the authors who have contributed and who realize the importance of this effort. We also wish to acknowledge many others who were able to attend the conference in Nepal but could not meet the deadline for contributions to this book. We hope that their contributions will appear in our future publications.

Anil Mital, Cincinnati, Ohio
Arun Pennathur, El Paso, Texas

June 2010

PART I

ENERGY

CHAPTER 1

SUSTAINABILITY AND THE CHALLENGE OF GOING GREEN

Shihab Asfour

Department of Industrial Engineering
University of Miami

Moataz Eltoukhy

Department of Industrial Engineering
University of Miami

PREFACE

One of the greatest challenges that will be facing the manufacturing industry in the years to come is the implementation of environmentally friendly manufacturing, not only from the engineering point of view, but also from a business perspective. This challenge requires manufacturers to give up their cynicism of the green manufacturing movement and to start seriously addressing the impact of their products on the environment.

It is evident that interest in green manufacturing does not always translate into the actions to implement the necessary changes. Thus it is important to provide some facts regarding green manufacturing to show the urgency of the implementation of such strategy, especially at a time when the pollution trend lines are alarming and the green manufacturing movement is becoming prominent.

This chapter will review the different aspects of sustainability and green manufacturing and will show how they can influence other aspects of manufacturing such as energy consumption, material selection, and so on. Also, some case studies will be discussed to show how organizations can go in the wrong direction when not considering going green.

INTRODUCTION

For years one main problem with going green is that it was believed that green approaches are more expensive than the traditional ones. As a result it was difficult to create a market for it.

According to a recent survey, manufacturers increasingly see going green, which can be achieved through both cost savings and enhanced efficiency, as an approach to move business forward. It was concluded that the cost barrier to green manufacturing is shrinking and that environmentally friendly practices can be successfully combined with traditional business practices (GreenBiz.com, 2008b).

A shift in acceptance of green manufacturing starts to be evident every day, which results in a serious interest from the manufacturing industry. On the other hand an interest in green manufacturing does not necessarily translate into a serious commitment to making the changes needed. Lanner Group, a business process analysis organization, found out that only 9% of survey respondents prioritized reduction of their carbon footprint in 2007 and 15% of the respondents said that the process is too costly (Atkinson, 2008).

There is no doubt that there is a strong belief among some manufacturers that going green costs money, causing these manufacturers to wait until they have extra money to cover the expenses of such a process. The truth is, with proper training in information technology (IT), manufacturers can identify a number of opportunities without any additional costs to them (Atkinson, 2008).

However, manufacturers who have invested money in green processes actually have been able to cut the cost of their products dramatically. As a result, this will provide the company with a competitive market advantage.

SUSTAINABILITY

Based on a United Nations definition, sustainability can be described as continual improvement of business operations to ensure long-term resource availability through environmental and transparent performance as it relates to consumers, business partners, and the community.

It might sound like understanding sustainability is an easy task, yet with conflicting consumer and customer demands, network complexities, and economic uncertainties, it is absolutely not.

Although consumers want greener packaging, they also want customized sizes of products, mixed pallets of goods, and reduced costs for the whole supply chain cycle. On top of that manufacturing issues such as inefficiency across the supply chain networks and the need for more direct-to-store deliveries add to the complexity of the problem.

Manufacturing Sustainability

Sustainable energy practices involve improving energy conservation and efficiency as well as utilizing alternative energy sources where possible. In manufacturing industries, a large portion of the total energy and water consumed and waste and emissions generated are due to production processes. New approaches to energy management are crucial to achieve substantial reductions in energy usage and emissions.

It is essential for society to limit the increase in earth's temperature while conserving its resources in order to avoid severe environmental consequences, which will require a climate-driven economy.

New technology advances may emerge in the future to support sustainable manufacturing. Meanwhile, progress can be made by taking a fresh look at many existing manufacturing, supply chain, plant engineering, and maintenance practices. It is also important for manufacturers to factor the involved energy, waste, and water burden associated with the existing facilities into their sustainability strategies.

Switching to a sustainable manufacturing model usually requires significant changes throughout the organization, including a change to manufacturing operations, design and engineering processes, and energy and utilities management, as well as an optimization of their process control.

According to the International Energy Agency (IEA) (2008), about one-third of the total global energy use and almost 22% of worldwide carbon dioxide emissions are due to the manufacturing industry. The IEA also identified some guidelines to increase the efficiency of certain fields in industry, such as the steel industry. Following those guidelines can result in an average 25% improvement in production using the existing technology. Some areas where plant departments can contribute to reduced energy consumption include the following:

- Improving the efficiency of industrial boilers, chillers, compressors, and so on
- Reducing steam and compressed air consumption
- Replacing older electric motors with more efficient and properly sized motors

Water Resources Sustainability

The concept of sustainable resources has been around for a long time. According to Loucks (2000), sustainable water resource systems are defined as "water resource systems designed and managed to fully contribute to the objectives of society, now and in the future, while maintaining their ecological, environmental, and hydrological integrity" (p. 30). Most definitions of sustainable water resources are so broad that they lack any quantitative definition. Dan Rothman defines sustainable water resources as "the ability to provide and manage water quantity and quality so as to meet the present needs of humans and environmental ecosystems, while not impairing the needs of future generations to do the same."

Water sustainability can also be defined as the ability to use water in sufficient quantities and quality from the local to the global scale to meet the needs of humans and ecosystems to sustain life for the short and long term and to protect humans from the damages brought about by natural- and human-caused disasters that affect sustaining life.

Sustainable water use has been defined by Gleick (1998) as "the use of water that supports the ability of human society to endure and flourish into the indefinite future without undermining the integrity of the hydrological cycle or the ecological systems that depend on it."

Because water impacts so many aspects of our existence, there are many facets that must be considered, such as the availability of freshwater supplies throughout periods of climatic change, the dangers of extended droughts, and the necessity to leave needed supplies for future generations. It also includes having the infrastructure for clean water and for treating water after it has been used by humans before being returned to water bodies.

Future Goals for Current Water Challenges

Current water challenges and future goals—which include the access to safe drinking water—are clearly stated in the "Millennium Development Declaration (2000)" and are expanded in the plan of implementation of the world summit on sustainable development. The plan of implementation outlines several central statements related to freshwater and sanitation issues such as the following (Mays, 2007):

- Develop integrated water resources management and water efficiency plans with support to developing countries.
- Support developing countries in their efforts to monitor and assess the quantity and quality of water resources.

GREEN MANUFACTURING

It appears that green manufacturing is becoming more of a reality. Many manufacturers are not only reducing their carbon footprint but also developing new products that are more environmentally friendly, as well as innovating items that will meet the sustainability goals of both companies and consumers.

Leading organizations like General Electric have engaged in this activity for some time now. Also, industries like the automotive industry are investing in green technologies, hoping that this will attract new customers for their products.

Adapting a green approach can be viewed as a process that uses a more eco-friendly supply that comes from renewable resources. These materials are to be used to produce products that meet sustainable product standards by organizations like the American National Standards Institute (ANSI) (Pojasek, 2008). Attention must also be paid to the direct and indirect use and loss of resources at the different product and service processes, which include the building and use of air compressors, boilers, and so on. Many manufacturing firms use a variety of management standards to effectively manage their processes for eco-efficiency.

The first place to start when it comes to turning a plant green is reducing energy use. In the long term, it is essential to develop green technology and products that take the manufacturing business to a higher level of green (Pojasek, 2008).

Green manufacturing reduces cost across the different aspects of the manufacturing process, including energy consumption, material selection, and material flow. Sustainability is now being added as another important factor to be considered in the decision making in manufacturing firms—at the same level of importance as both cost and quality.

A good definition of sustainability comes from the United Nations: "Sustainable development is development that meets the needs of the present without compromising the ability of future generations to meet their own needs" (Brundtland Commission, 1987).

Some sectors are interested in technologies that eliminate harmful materials, such as lead, from products during processing. In the automotive industry, for example, the focus is on technology that can reduce emissions of harmful toxins as well as finding replacement materials that improve conservation efforts.

There is no doubt that interest in sustainability issues has recently increased dramatically. One of the explanations of what is causing this ramped-up interest in sustainability is simply the desire to "do the right thing"; another factor coincides with the public's increasing awareness of green issues, aided by widespread environmental information and reports. For instance, most book publishers like the idea that their supply chain partners are offering green choices. And the ability to offer greener choices relies on whether these publishers are engaged in sustainability issues.

One example of this kind of effort is the Green Edition Labeling Program, conducted by Courier. This program offers a "Green Edition" label to publishers whose books are printed on recycled paper and printed in the United States, which results in reducing the environmental impacts of distribution (Rowzie & Ynostroza, 2008).

The use of the simulation technology in resource planning and material delivery logistics also provides another useful tool. In fact, according to the Lanner Group's survey, 90% of the respondents say that simulation technology can facilitate greener manufacturing provided that its potential is maximized. Simulation can provide visibility and clarity and eliminate unexpected costs that are common in new projects. Also, it can map processes so that environmental initiatives can be aligned with cost savings (Rowzie & Ynostroza, 2008).

CASE STUDIES

The main goal of green manufacturing, which relies on minimizing both waste and pollution, is to support future generations by attaining sustainability, and that is done mainly by preserving natural resources. This is basically done through product process and design by designing the product or process more efficiently.

Although the logistics industry is accused of being a major contributor to the global warming problems, in reality, it is just a very visible industry to the general public. For instance, in the food supply chain, transportation is responsible for just 2.5% to 3.5% of total UK greenhouse gas emissions, while the meat and dairy industry generates 8%.

Traditionally, public opinion as well as the media associate logistics with high greenhouse gas emissions. However, recent carbon footprint life cycle assessments on products such as beer, chips, and smoothies showed that the distribution element of the total footprint was only about 10%.

There are two different approaches to environmental issues. The first approach is adapted by companies that do not spend time on the process of sustainability; they only use information they already gathered for their own management requirements. The second approach is the one utilized by companies that spend time gathering information to develop their own database that can be used in identifying changes in each area of the supply chain.

FedEx Corporation

FedEx has already reduced aircraft carbon dioxide emissions by 3.7% per available ton-mile and improved vehicle fuel efficiency by 13.7%, reducing vehicle carbon emissions by almost one billion pounds since 2005 (GreenBiz.com, 2008a).

The company plans to reduce 20% of the carbon dioxide emissions from its aircraft fleet and also improve the fuel efficiency of its vehicle fleet by 20% by the year 2020 (GreenBiz.com, 2008a).

In order to achieve its fuel efficiency goals, FedEx has invested in more fuel-efficient ways of operating by doing the following:

- Operating the largest fleet of commercial hybrid electric trucks in North America, which resulted in an improvement in fuel economy by 42% and a reduction of greenhouse gas emissions by approximately 25%.
- Converting one quarter of its fleet to more fuel-efficient vehicles, saving more than 45 million gallons of fuel in the last 3 years.
- Upgrading its aircraft fleet by replacing Boeing 727 aircraft with Boeing 757 planes, which reduce the environmental impact, cutting fuel consumption by up to 36% while providing 20% more space.

UPS

UPS has ordered 12 zero-emission electric vehicles—six for deployment in Germany and six for the United Kingdom. Those vehicles are the first of their kind to effectively control the power of modern, high-energy batteries to meet the high performance requirements of urban delivery vehicles.

UPS currently operates the transportation industry's largest private fleet of alternative-fuel vehicles, which are deployed in a number of countries such as the United Kingdom, the United States, Brazil, Canada, France, Germany, and Mexico. It is worth mentioning that UPS began using alternative-fuel vehicles in New York City in the 1930s.

Textiles and Plastics

A University of Alabama chemist has invented a new way to dissolve and use cellulose, found in the cell walls of trees, in producing environmentally friendly materials that have potential for industries such as automotive, packaging, and textile (Bryant, 2005).

This new technique is based on the discovery of a specific solvent, known as 1-butyl-3-methylimidazolium chloride, which dissolves cellulose. This liquid is part of a new class of solvents, known as ionic liquids, which are typically nontoxic, nonflammable, and do not evaporate, thus significantly reducing harmful emissions.

Cellulose-based plastics, which have been used on a limited basis, are not as popular as petroleum-based plastics, such as polyethylene, which are commonly used in soft drink bottles and milk jugs. Petroleum-based plastics are not biodegradable, and they often accumulate in landfills.

SUMMARY

Going green is the transformation of moving into an environment-friendly enterprise that delivers benefits that can affect the company, the consumer, and their communities (BusinessKnowledgeSource.com, 2007). This can be achieved by using machines, tools, and resources that are eco-friendly, such as wind power. A totally green manufacturing plant will benefit from renewable and reusable resources and the cost savings therein.

The major benefit of going green is that the changes that occur can significantly impact the damage that is done to the environment and the earth. Despite the fact that many opportunities to cut emissions in the supply chain pay for themselves over time, most companies still do not consider emission reduction in their strategic planning (BusinessKnowledge Source.com, 2007).

The efforts for energy efficiency and production that is not harmful to the environment will continue. Those efforts have come a long way, yet many are looking forward to the possibilities that exist for future generations of manufacturers. Collaboration with customers and suppliers remains essential in the achievement of those environmental goals.

The three main recommendations to be adopted when changing a manufacturing plant to green are (a) to design a product to be reused, which means to design products that can be used in later generations of products; (b) to develop products that can be taken apart easily; and (c) to develop a method for manufacturing products in a way that the parts can be used in different products.

Also, the simplest way to start when it comes to changing a manufacturing plant to green is by reducing energy consumption. In addition, by incorporating environment-friendly ambitions with long-term goals of decision makers as well as promoting the development of products, the manufacturing plant will attain the sustainable level of green needed.

It is of a great importance also to always remember that the industrial green process involves smart design of all products, processes, systems, and organizations, as well as the implementation of smart strategies that harness both the technologies and the ideas in an effective way that prevents any environmental problems before they even arise (Richards, 1997).

Finally, as stated in Townsend (1998), "Unless we change we'll get where we're going." The realization of the facts behind this statement in relation to the environment is what has motivated people to examine their behavior and try to find new paths.

REFERENCES

Atkinson, W. (2008). Green manufacturing: Production can be environmentally friendly and cost-effective. *Manufacturing Business Technology*. Retrieved from http://www.rit.edu/news/dateline_archive/datelinerit_apr03_2008.html#20767908

Bryant, C. (2005). UA's Rogers wins presidential green chemistry challenge for enviro-friendly method of producing textiles and plastics. *University of Alabama News*. Retrieved from http://uanews.ua.edu/anews2005/jun05/greenchem062005.htm

BusinessKnowledgeSource.com. (2007). The benefit of going green in your manufacturing. Retrieved from http://www.lohas.com/articles/100810.html

GreenBiz.com. (2008a). FedEx, IBM, and Office Depot report green progress. Retrieved from http://www.greenbiz.com/news/2008/11/13/fedex-ibm-and-office-depot-report-green-progress

GreenBiz.com. (2008b). Green manufacturing can help "move business forward." Retrieved from http://www.greenbiz.com/news/2008/08/15/green-manufacturing-can-help-move-business-forward

International Energy Agency (2008). Worldwide trends in energy use and efficiency: Key insights from IEA indicator analysis. Retrieved from http://www.iea.org/papers/2008/indicators_2008.pdf

Loucks, D. P., & Gladwell, J. S. (2000). *Sustainability criteria for water resource systems*. Cambridge, UK: Cambridge University Press.

Mays, L. W. (2007). *Water resources sustainability*. New York, NY: McGraw-Hill.

Pojasek, R. B. (2008). When is green manufacturing green? SMEspeaks. Retrieved from http://www.robertpojasek.com/uploads/SME-speaks_-_81408_.pdf

Richards, D. J. (1997). *The industrial green game*. Washington: National Academy Press.

Rowzie, K., & Ynostroza, R. (2008). Green manufacturing gains momentum. *Publishers Weekly*. Retrieved from http://www.publishersweekly.com/article/CA6616667.html

PART II

MATERIALS

CHAPTER 2

ENHANCING RESOURCE UTILIZATION THROUGH THE DEVELOPMENT OF A DISASSEMBLY ALGORITHM

Anoop Desai

Department of Mechanical and Electrical Engineering Technology
Georgia Southern University

Anil Mital

Department of Mechanical Engineering
University of Cincinnati

INTRODUCTION

It has now been well established that the earth's natural resources are finite. There has been considerable research in different branches of science and technology that has addressed various kinds of complex problems in bygone years. Scientists have explored various phenomena within physical, biological, and chemical science. They have formed algorithms to solve complex problems within the engineering discipline and its various subcategories such as mechanical engineering, electrical engineering, chemical engineering, and so on. This research and the subsequent solutions produced from it have served mankind extremely well over the years. However, we are now faced with a very different kind of problem, one that cannot be easily solved. It cannot even be easily approached. The problem concerns natural resources.

If you were to give it careful thought, all products that are manufactured in whatever shape, size, or form can be traced back to some naturally occurring element that can be found on earth. For instance, glass can be traced back to silicon, plastic can be traced back to crude oil, and wood can be traced back to trees. Each and every source of material can be found in the natural environment. As a matter of fact, in spite of all the research that has gone into bettering the human condition, not a single effort has been successful in creating a completely artificial element from the ground up.

The basic takeaway from this discussion, therefore, is that in the broad scheme of things, human civilization is a function of its natural environment. We depend on natural resources for our livelihoods, we depend on natural resources to drive our technologies, no matter how advanced they may be, and we depend on natural resources for our very existence.

The earth has a limited supply of natural resources. For instance, of all the coal that was mined over the last 800 years, over 50% was mined in the last 40 years. Now that is a statistic that, if not frightening, is certainly mind boggling. It causes you to sit up and take notice. Humanity needs to start asking, What happens if we run out of natural resources? How would we survive? For instance, if we run out of crude oil, how do we satisfy our thirst for plastic products? These are questions that are not easily addressed.

Recycling, Reuse, and Remanufacturing as a Means for Optimizing Use of Material Resources

Recognition of the scarcity of natural resources and acceptance of this concept as a fact have led to the dawning of a new era of environmental consciousness. This has lent credence to end-of-life (EOL) objectives such as component reuse (components from a retired product being used in a new product without being upgraded), component remanufacturing (components from a retired product being used in a new product after upgrading it technologically), and component recycling (reuse at the material level, e.g., recycling of plastics), which constitute some of the most important reasons for disassembling products. Widespread diffusion of consumer goods and shortening of product life cycles have led to an unprecedented number of used products being discarded. In 1991, Carnegie Mellon University estimated that by 2012, some 150 million obsolete PCs, none with readily recoverable materials, will require more than 8 million cubic meters of landfill space at a cost of around $400 million (Lee & Khoo, 2001). However, the number of potential landfill sites for nonhazardous solid wastes has seen an exponential decrease. In the United States alone, landfill sites have diminished from 18,000 in 1985 to 9,000 in 1989. According to a recent study, the United States had lost more than 70% of its landfill sites by 1997 (Zhang, Kuo, Lu, & Huang, 1997), with landfills in many states reaching their permitted capacities at an alarming rate.

The reclaiming approach seeks to offer a solution to optimize material use. This approach utilizes material from the supply chain from products that have been retired from active life and reintroduces them at the beginning of the supply chain. There are various forms of reclaiming that are presented in Table 2.1. The following paragraphs will introduce the basic concepts of recycling, reuse, and remanufacturing as a means for optimization of material use.

The different kinds of reclamation are a function of the objective (e.g., the objective could be stated as "to restore working condition" or "to restore quality level as new").

Table 2.1. Product Recovery Options After Disassembly

Options	*Objective*	*Level of disassembly*	*Result*
Repair	Restore working condition	Product level (limited disassembly and fixing)	Some parts repaired
Refurbishing	Improve quality level (though not like new)	Module level (some technological upgrading)	Some modules repaired or replaced
Remanufacturing	Restore quality level as new	Part level	Used and new parts in new products
Cannibalization	Limited recovery	Selective disassembly and inspection of potentially reusable parts	Parts reused or recycled and disposed of
Recycling	Reuse materials only	Material level	Materials used in new products

Source: Thierry, Salomon, van Nunen, and van Wassenhowe (1995).

Recycling and Reuse to Optimize Material Use

It is clear from Table 2.1 that recycling is done at the basic material level for a product. Recycling turns materials that would otherwise become waste into valuable resources. Collecting used bottles, cans, and newspapers and taking them to the curb or to a collection facility is just the first in a series of steps that generates a host of financial, environmental, and social returns. Some of these benefits accrue locally as well as globally.

Recycling can be classified into four different categories, namely primary recycling, secondary recycling, tertiary recycling, and quaternary recycling. The level of recycling is a function of the composition of the element being recycled as is evident from Table 2.2.

The benefits of recycling are enumerated as follows (EPA, 2009):

- Recycling reduces the need for landfills and incineration.
- Recycling prevents pollution caused by the manufacturing of products from virgin materials.
- Recycling saves energy.

Table 2.2. Type of Recycling According to Component Composition

Adaptation of recycling to composition of component		
Type of recycling	*Definition*	*Composition of the component*
Primary recycling	Recycling on a comparable quality level	No alloy present in the component; polymer content in the component
Secondary recycling	Recycling on a lower quality level; *down cycling*	Presence of an alloy in the component; no polymer content; ceramic content; elastomer or composite material
Tertiary recycling	Decomposition	
Quaternary recycling	Incineration with energy retrieval	No polymer content; ceramic content

- Recycling decreases emissions of greenhouse gases that contribute to global climate change.
- Recycling conserves natural resources such as timber, water, and minerals.
- Recycling helps sustain the environment for future generations.

As its name suggests, the process of reuse involves using a product more than just once. This includes conventional reuse in which the item is used again for the same function and new life reuse in which it is used for a new function. This can be contrasted with recycling wherein a used item is broken down into raw materials, which are used to make new items. By taking useful products and exchanging them without reprocessing, reuse helps us save time, money, energy, and resources. In broader economic terms, reuse offers quality products to people and organizations with limited means while generating jobs and business activity that contribute to the economy.

Historically, financial motivation was one of the main drivers of reuse. In the developing world this driver can lead to very high levels of reuse; however, rising wages and consequent consumer demands for the convenience of disposable products have made the reuse of low-value items such as packaging uneconomic in richer countries, leading to the demise of many reuse programs. Current environmental awareness is gradually changing attitudes and regulations, such as the new packaging regulations, and is gradually beginning to reverse the situation.

Potential advantages of reusing a product are presented in the following list:

- Saves energy and raw materials, as replacing many single-use products with a reusable one reduces the number that need to be manufactured
- Reduces disposal needs and costs

- Brings sophisticated, sustainable, well-paid jobs to underdeveloped economies
- Saves money for businesses and consumers, as a reusable product is often cheaper than the many single-use products it replaces

Remanufacturing to Optimize Material Use

Remanufacturing is defined as the process of disassembly and recovery at the module level and, eventually, at the component level. It requires the repair or replacement of worn out or obsolete components and modules. Parts subject to degradation affecting the performance or the expected life of the whole are replaced. Remanufacturing differs from other recovery processes in its completeness: A remanufactured machine should match the same customer expectation as new machines.

Differences Between Recycling and Remanufacturing

Remanufacturing and recycling are two different concepts. However, many environmental groups are embracing the idea of remanufacturing over recycling because it cuts down on the use of energy and resources used for processing. Recycled goods are consumed and then returned to their original raw material form to be used again. Remanufacturing, on the other hand, recycles the value originally added to the raw material.

Remanufacturing differs from recycling because it makes a much greater economic contribution per unit of product than recycling does. The basic difference arises in the recapture of value added. Value added is the cost of labor, energy, and manufacturing operations that are added to the basic cost of raw materials in the manufacture of a product. For all but the simplest durable goods, value added is by far the largest element of cost. In a product such as a beer bottle, the cost of the basic raw materials (sand, soda, and lime) is much less than 5% of the cost of a finished bottle. The rest is value added. For a product such as an automobile, the value of the raw materials that can be recovered by recycling is only in the order of 1.5% of the market value of the new car. Value added is embodied in the product. Recycling destroys that value added, reducing a product to its elemental value—its recoverable raw material constituents. Further, recycling requires added labor, energy, and processing capital to recover the raw materials. When all of the costs of segregation, collection, processing, and refining are taken into account, recycling has significant societal cost. This forms the most elemental point of difference between recycling and remanufacturing. However, as the reader can easily appreciate, it is of immense economic and therefore societal value.

Product Disassembly to Facilitate Recycling, Reuse, and Remanufacturing

The gravity of the situation as outlined in the preceding sections has prompted increasing governmental pressure on manufacturers in the form of various environmental laws that have been drafted to obligate industry to continue assuming responsibility for their products even after their useful life. Products will be expected to derive minimal energy and resources from the environment and discharge a minimal amount of waste during and after their life cycles. For example, the German policy on combating waste from EOL vehicles requires car manufacturers to take back EOL vehicles free of charge from the last owner in order to enable material recycling. Unrecyclable material is required to be disposed of without causing any harm to the environment or to human health. Disassembling products for EOL objectives allows reusable, nonrecyclable, and hazardous subassemblies to be selectively separated from recyclable ones.

From the manufacturers' or reclaimers' perspective, the process may be clearly distinguished into two categories based on the method of disassembly:

- Destructive disassembly or brute force approach, for example, incineration, metal cutting, and so on
- Nondestructive disassembly or reverse assembly

In the case of nondestructive disassembly, if a fastener is screwed in, then it is screwed out. If two parts are snap-fit together, then they are snapped out. In the case of destructive disassembly, parts are just pulled or cut (Kuo, Huang, & Zhang, 2001).

Depending on the extent of disassembly, nondestructive disassembly can be further classified into two categories as follows:

- *Total disassembly*. The entire product is disassembled into its constituent components. This is sometimes not economically feasible due to the imposition of external constraints such as time, economic factors, presence of hazardous materials, and so on.
- *Selective disassembly*. Selective disassembly is defined as the reversible dismantling of complex products into less complex subassemblies or single parts (Lambert, 1999). It involves the systematic removal of desirable constituent parts from an assembly while ensuring that there is no impairment of parts due to the process (Brennan, Gupta, & Taleb, 1994). Products are selectively disassembled in order to realize the following objectives:

    1. Enabling maintenance and repair activities (serviceability). This enables the ease of performing all service-related operations including maintenance, malfunction, diagnosis, and repair. Benefits of designing for serviceability include reducing warranty costs (higher earnings before interest and tax).
    2. Enhancing customer appeal and lengthening the service life due to product ability to be serviced economically.
    3. Availability of subassemblies as service parts or for assembly in new products. Components that are still robust enough for extended use in another product are reused to achieve economies of scale by component reuse (cost optimization).

4. Availability of parts intended for material reuse (recycling).
5. Complying with regulations that prescribe removal of definite parts, materials, and substances for environmental and safety reasons, for example, removal of working fluids such as engine oils, lubricants, and so on.

AN OVERVIEW OF THE RESEARCH ON DESIGN FOR DISASSEMBLY

The following section offers a brief glimpse into the various strategies that have been adopted by researchers and scientists to enable efficient material retrieval and thus try to optimize material utilization.

The bulk of research conducted on disassembly has concentrated on such issues as disassembly sequence planning, disassembly evaluation and analysis, and product recovery. Thierry, Salomon, van Nunen, and van Wassenhowe (1995) proposed a product recovery management (PRM) approach wherein returned products can be recovered at four levels: product, module, part, and material level (in that order). The objective of this methodology is to recover as much as possible of the economical as well as ecological value of products, components, and materials so as to reduce the ultimate quantities of waste to a minimum.

Krikke, van Harten, and Schuur (1998) considered the problem at the tactical management level to *determine an optimal product recovery and disposal (PRD) strategy*. The model uses the disassembly tree as the starting point to describe the disassembly process for the return product. Retrievable parts, modules, and subparts are identified and represented at various sublevels. The aim of disassembly is to make separate recovery or disposal possible for every single subassembly. In this particular approach, materials are not considered as the lowest disassembly level.

A disassembly sequence plan (DSP) is a sequence of disassembly tasks that begins with a product to be disassembled and terminates in a state where *all the desired parts of the product are disconnected* (Gungor & Gupta, 1998). A DSP aims to optimize product recovery through the minimization of cost, maximization of material recovered, and minimization of disassembly time using mathematical techniques such as linear programming, dynamic programming, and graphical tools.

The problem of dealing with future uncertainties of recycling options is addressed by Zussman, Kriwet, and Seliger (1994), who use the utility theory. Attributes such as technology refinement, prices, or dumping fees are bound to change with time (dynamic) and as such involve uncertainty for the designer or policy maker. The methodology incorporates the optimization of disassembly sequences to achieve objectives such as maximization of net profit, maximization of parts reuse, or minimization of waste headed toward landfills. A *recovery graph* is constructed using the "and/or" logic as described earlier. Values of different entities such as disassembly sequences, recycling options, and related costs are used to plot the *recovery graph*. The principal drawback of this approach is it does not specify the time horizon over which future uncertainties are being planned.

Two algorithms were presented by Gungor and Gupta (1997) to obtain a scheme for disassembling multiple product structures that have common parts. The core algorithm determines the number of root items to be disassembled in order to minimize disassembly cost. A root is defined as a major subassembly that is further composed of minor subassemblies that can be subdivided into individual components. The allocation algorithm is used to determine a disassembly schedule for the roots and subassemblies by allocating the disassembly requirements over the planning horizon.

Kuo (1999) divided disassembly planning into four distinct stages: geometric assembly representation, cut-vertex search analysis, disassembly precedence matrix analysis, and disassembly sequence and plan generation for obtaining the disassembly sequence of electromechanical products during the design stage. The disassembly cost is classified as target disassembly, full disassembly, and optimal disassembly. The disassembly tree analysis is the basic idea behind disassembly sequence generation. Disassembly precedence matrices and graphs that depict the relationship between components and fasteners represent interrelationships among components. The disassembly process is simplified by forming disassembly modules. The cut-vertex search analysis enables the decomposition of a module into submodules. A cut-vertex in a component-fastener graph is a vertex (component) whose removal disconnects the graph. In a real component, a cut-vertex is the main part that is connected to other components. In a module, a cut-vertex is the connection component between two other components. Three distinct types of geometric assemblies can be defined:

- Type I. An assembly having a main component to which other components or subassemblies are directly or indirectly assembled.
- Type II. An assembly having no main component. All components are assembled with others. This can be disassembled only as a single component.
- Type III. An assembly that is a combination of both of the earlier types. This can be disassembled into multiple components.

Three kinds of disassembly sequences driven by disassembly cost are analyzed upon formation of the disassembly tree.

- Type I (target disassembly sequence [TDS]). Specific components are disassembled in order to remove valuable components.
- Type II (optimal disassembly sequence [ODS]). Disassembly is stopped when marginal return on the operation becomes uneconomical.
- Type III (complete disassembly sequence [CDS]). Complete disassembly of the product.

The more important economic considerations to be taken into account during the disassembly process include such factors as value added to products and materials during manufacturing, disassembly cost and revenue per operation, and the penalty if poisonous materials are not completely removed (De Ron & Penev, 1995). Operating costs continue to be one of the most daunting concerns for manufacturers. The EOL economic value of components can be computed using the costing technique suggested by Lee and Khoo (2001). This technique employs conventional costing practices in addition to specifying a miscellaneous cost (the summation of collecting cost and processing cost) that is in turn used to calculate values of other entities such as reuse value, remanufacture value, primary and secondary recycling values, incineration value, and landfill cost. The role played by economic factors in the determination of an optimal recycling and disassembly strategy is illustrated in Figure 2.1.

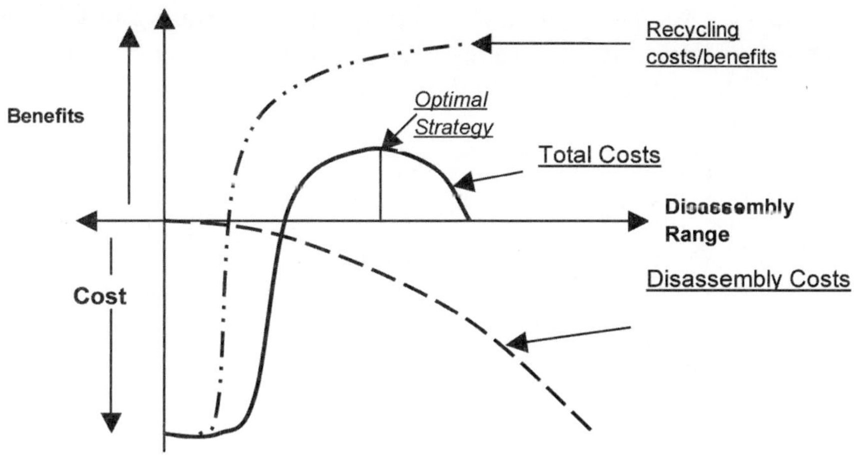

Figure 2.1. Determination of optimal recycling and disassembly strategy.
Source: Adapted from Feldman et al. (1999).

A NOVEL METHODOLOGY FOR PRODUCT DISASSEMBLY TO OPTIMIZE MATERIAL USE

The preceding section offered a brief overview of the different kinds of approaches adopted by various scientists globally to address the problem of effective and efficient product disassembly.

However, it has to be noted that disassembly is a practical problem. No amount of mathematical formulation can serve to solve what is essentially a labor-oriented problem. Over 70% of product life cycle costs are ascertained at the design stage itself. This makes it important to have a methodology that seeks to accomplish proactive product design for disassembly.

A practically applicable methodology to enhance efficient disassembly of products will be presented in this section. This disassembly methodology takes into consideration numerous factors such as the weight, size, and shape of components being disassembled; frequency of disassembly tasks (based on number of similar products being disassembled within a particular time frame); requirement of manpower; postural requirements; material handling requirements; and need for component preparation such as cleaning and degreasing tasks. A number of human factors in addition to design and economic factors merit consideration due to the labor-intensive nature of the disassembly process. These factors directly affect the disassembly process and had hitherto been neglected in the formulation of both disassembly algorithms and Design for Disassembly (DfD) methodologies.

The most common and widely used disassembly operations are recorded and described in sufficient detail. Every disassembly operation is then subdivided into basic elemental tasks. It has been observed that only a fraction of all the tasks in the entire disassembly operation are actually responsible for performing disassembly. The remaining tasks constitute such actions as reaching for tools, grasping tools, and cleaning components prior to disassembly.

The new methodology consists of the following distinct elements:

- A numeric disassemblability evaluation index
- Systematic application of DfD methodology

The numeric disassemblability evaluation index is a function of several design parameters that directly or indirectly affect the process of consumer product disassembly. Numerical scores are assigned to each of these parameters depending on the ease with which they can be attained. The following parameters have been addressed:

- *Degree of accessibility of components and fasteners.* Easy access is a prerequisite for quick and efficient disassembly operation. The less accessible a component or fastener is, the higher numerical score it receives.
- *Amount of force (or torque) required for disengaging components (in case of snap-fits) or unfastening fasteners.* The lesser the amount of force required, the better the design. The amount of effort required is directly proportional to the value of the numerical score received.
- *Postural requirements for performing disassembly tasks.* The disassembly process is still predominantly labor intensive and, according to a recent Delphi study, is expected to remain so in the foreseeable future. As a result, certain disassembly operations that require the assumption of unnatural postures would prove to be highly detrimental to the operator performing those operations. An unnatural posture can be defined as one responsible for the onset of static muscular fatigue. This issue assumes even greater importance in the light of high frequency of disassembly tasks. A provision for including additional allowances in the disassembly score based on this category has been made in the methodology.
- *Identification of dirt traps.* This factor is important for obvious reasons. A product that has been in regular use is bound to accumulate internal dirt over a period of time. From the disassembly perspective, components that accumulate dirt need to be cleaned and degreased before disassembly and therefore involve prior preparation. This is a

time- and labor-intensive activity. Empirical data can enable easy identification of dirt traps at the design stage. This can enable component redesign to facilitate disassembly.

- *Design factors such as weight, shape, and size of components being disassembled.* This can be a crucial consideration in product disassembly, especially since it involves the use of special fixtures and apparatuses or simply more manpower. For example, the CRT of a 25-inch television set can be quite heavy and large for a single person to manipulate efficiently. These factors have been addressed through the introduction of an additional multiplier for material handling.

Identification of design anomalies helps optimize component design from the disassembly perspective. Several design factors such as accessibility, mating surface condition, corrosion, size, weight, shape, and so on play an important role in disassembly. It is therefore imperative that they should be addressed in detail before a particular component design is finalized. The application of the disassembly evaluation criteria to a product results in numerical indices for various categories of evaluation. These scores can be multiplied by additional allowances prespecified for posture, motions, visual fatigue, and manpower requirements. The higher the score an evaluation category obtains, the greater the chance of detecting a design flaw within that category.

An example of numeric scores to be assigned for disassembly force is presented in Table 2.3. The simplest disassembly task of removing an easily grasped object without the exertion of much force by hand by a trained worker under average conditions has been considered as the basic disassembly task. A score of 73 time measurement units (TMUs) was assigned to this task, which corresponded to time duration of approximately 2 seconds. Subsequent scores were assigned based on the detailed study of the most commonly encountered disassembly operations.

Allowances for various attributes that affect the dismantling process have been presented in Table 2.4. Relevant allowances include those made for posture, manpower, fatigue, and types of motions.

Figure 2.2 is a hierarchical representation of the DfD methodology being discussed.

ILLUSTRATIVE CASE STUDY

This section presents an illustrative case study to demonstrate the practical effectiveness of the methodology. The case of a computer monitor disassembly is presented. Table 2.5 indicates the various components used in a computer CRT.

CONCLUSION

This chapter emphasized the importance of conserving natural resources for the betterment of mankind. One of the approaches that would enable the achievement of this objective is to reuse, recycle, or remanufacture material and components at the end of their useful lives. In

Table 2.3. Numeric Scores Assigned to Varying Amounts of Disassembly Force

Design attribute	*Design feature*	*Design parameters*	*Score*	*Interpretation*
Disassembly force	Straight line motion without exertion of pressure	Push/pull operations with hand	0.5	Little effort required
			1	Moderate effort required
			3	Large amount of effort required
	Straight line and twisting motion without pressure	Twisting and push/pull operations with hand	1	Little effort required
			2	Moderate effort required
			4	Large amount of effort required
	Straight line motion with exertion of pressure	Intersurface friction and wedging	2.5	Little effort required
			3	Moderate effort required
			5	Large amount of effort required
	Straight line and twisting motions with exertion of pressure	Intersurface friction and wedging	3	Little effort required
			3.5	Moderate effort required
			5.5	Large amount of effort required
	Twisting motions with pressure exertion	Material stiffness	3	Little effort required
			4.5	Moderate effort required
			6.5	Large amount of effort required

Table 2.4. Postural Allowances for Unnatural Postures Adopted During Disassembly

Postural allowances	*Percentage multipliers (%)*
Sitting down	0
Standing up	2
Bending down	5
Lying down	3
Crouching	5
Stretching	8
Squatting	8

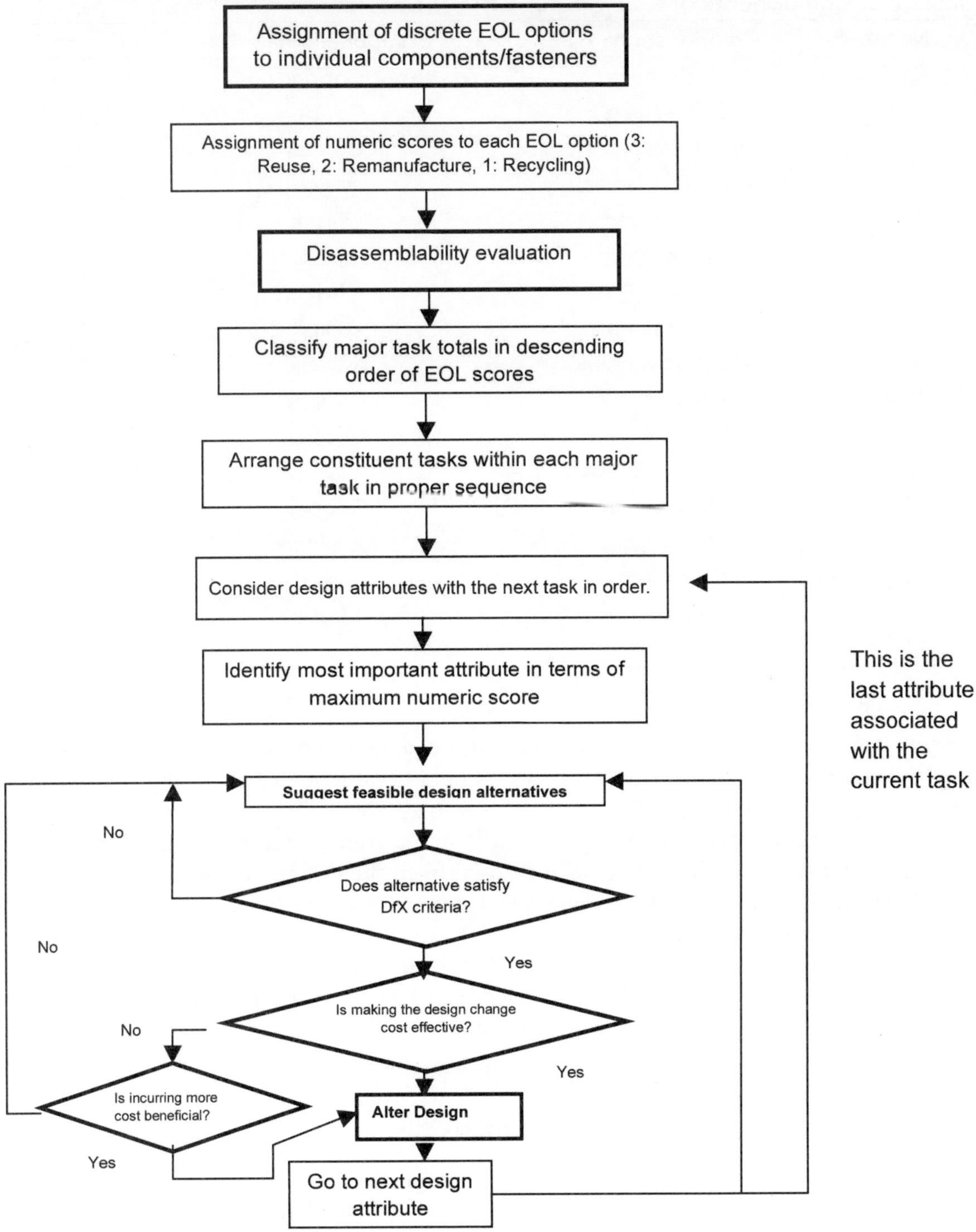

Figure 2.2. Decision-making hierarchy of DfD methodology.

Table 2.5. Components of a Computer CRT

No	*Component name*	*Component material*	*Quantity*
1	Back screw	Copper	4
2	PCB screws	Copper	2
3	CRT screws	Copper	4
4	CRT and PCB assembly	Mixed	1
5	Back cover	Plastic	1
6	Swivel base	Plastic	1
7	Pivot	Plastic	1
8	Yoke assembly	Mixed	1
9	Deflection wire lead	Mixed	1
10	Retainer screws	Copper	2
11	Main wire lead	Copper	1
12	Adjusting knobs	Plastic	4
13	PCB retainer screw	Copper	1
14	Retaining lugs	Aluminum	4
15	PCB assembly	Mixed	1
16	Rear board	Plastic	1
17	CRT	Mixed	1

order to be able to accomplish this effectively, it is essential to be able to take apart products that have reached the end of their useful lives. The process that enables taking apart product components is termed disassembly.

Various kinds of disassembly approaches were introduced in this chapter. A methodology that is practically applicable and can be used from the design perspective was also presented. The methodology is largely time based and hence can be used objectively on several different product structures.

Time is only one of the metrics that can be considered in evaluating the ease with which either assembly or disassembly can be carried out. Examples of other metrics that can be used in this evaluation are effort, cost, use of fixtures and equipment, workforce, and so on.

REFERENCES

Brennan, L., Gupta, S., & Taleb, K. (1994). Operations planning issues in an assembly/disassembly environment. *International Journal of Operations and Production Management*, *14*(9), 57–67.

De Ron, A., & Penev, K. (1995). Disassembly and recycling of electronic consumer products: An overview. *Technovation*, *15*(6), 363–374.

Table 2.6. Disassembling a Computer Monitor

Task number	Task description *Nondestructive disassembly of computer monitor*	Task total	Intersurface friction	Intersurface wedging	Material stiffness	Component size	Component weight	Component symmetry	Force exertion	Torque exertion	Dimensions	Location	Accuracy of tool placement	Posture allowance	Motions allowance	Manpower allowance	Visual fatigue allowance (%)
			(Dis)assembly force			Material handling			Tooling		Accessibility and positioning			Allowances			
1a	Rotate swivel base about pivot	13.54	–	4	–	2.0	2.0	1.2	1	–	1.0	1	1.2	–	–	–	1
1b	Pull out swivel base	11.51	–	2	–	2.0	2.0	1.2	1	–	1.0	1	1.2	–	–	–	1
2	Disassemble back cover																
2a	Unscrew 1st of four back screws	15.65	2.5	–	–	2.0	2.0	0.8	–	2	1.6	2	2.0	–	–	–	5
2b	Unscrew 2nd of four back screws	15.65	2.5	–	–	2.0	2.0	0.8	–	2	1.6	2	2.0	–	–	–	5
2c	Unscrew 3rd of four back screws	15.65	2.5	–	–	2.0	2.0	0.8	–	2	1.6	2	2.0	–	–	–	5
2d	Unscrew 4th of four back screws	15.65	2.5	–	–	2.0	2.0	0.8	–	2	1.6	2	2.0	–	–	–	5
2e	Remove back cover	15.44	–	3	–	3.5	2.0	1.2	–	1	1.0	1	2.0	–	–	–	5
3	Disassemble CRT/PCB assembly																
3a	Unscrew 1st of two PCB screws	15.65	2.5	–	–	2.0	2.0	0.8	–	2	1.6	2	2.0	–	–	–	5
3b	Unscrew 2nd of two PCB screws	15.65	2.5	–	–	2.0	2.0	0.8	–	2	1.6	2	2.0	–	–	–	5

5b	Remove 2nd retainer screw	15.65	2.5	–	–	2.0	2.0	0.8	–	2	1.6	2	2.0	–	–	–	5
5c	Remove main wire lead	13.13	–	3	–	4.0	2.5	2.2	1	–	1.6	1	2.0	–	–	–	1
6	Remove adjusting knobs																
6a	Remove 1st of four adjusting knobs	11.0	–	2	–	2.0	2.0	0.8	1	–	1.0	1	1.2	–	–	–	–
6b	Remove 2nd of four adjusting knobs	11.0	–	2	–	2.0	2.0	0.8	1	–	1.0	1	1.2	–	–	–	–
6c	Remove 3rd of four adjusting knobs	11.0	–	2	–	2.0	2.0	0.8	1	–	1.0	1	1.2	–	–	–	–
6d	Remove 4th of four adjusting knobs	11.0	–	2	–	2.0	2.0	0.8	1	–	1.0	1	1.2	–	–	–	–
7	Remove PCB assembly																
7a	Remove PCB retainer screw	18.27	5.0	–	–	4.0	2.0	0.8	1	–	1.6	1	2.0	–	–	–	5
7b	Bend 1st retaining lug	25.08	–	–	6.5	4.0	2.0	0.8	3	–	2.0	2	2.5	–	–	–	10
7c	Bend 2nd retaining lug	25.08	–	–	6.5	4.0	2.0	0.8	3	–	2.0	2	2.5	–	–	–	10
7d	Bend 3rd retaining lug	25.08	–	–	6.5	4.0	2.0	0.8	3	–	2.0	2	2.5	–	–	–	10
7e	Bend 4th retaining lug	25.08	–	–	6.5	4.0	2.0	0.8	3	–	2.0	2	2.5	–	–	–	10
7f	Remove PCB assembly	11.55	–	2	–	2.0	2.0	0.8	1	–	1.0	1	1.2	–	–	–	5
8	Remove swivel pivot																
8a	Pry out swivel support	18.7	–	–	4.5	3.5	2.0	1.2	2	–	1.6	1	2.0	–	–	–	5

Table 2.6. (continued)

Task number	Task description *Nondestructive disassembly of computer monitor*	Task total	Intersurface friction	Intersurface wedging	Material stiffness	Component size	Component weight	Component symmetry	Force exertion	Torque exertion	Dimensions	Location	Accuracy of tool placement	Posture allowance	Motions allowance	Manpower allowance	Visual fatigue allowance (%)
			(Dis)assembly force			Material handling			Tooling		Accessibility and positioning			Allowances			
3c	Unscrew 1st of four CRT screws	15.65	2.5	–	–	2.0	2.0	0.8	–	2	1.6	2	2.0	–	–	–	5
3d	Unscrew 2nd of four CRT screws	15.65	2.5	–	–	2.0	2.0	0.8	–	2	1.6	2	2.0	–	–	–	5
3e	Unscrew 3rd of four CRT screws	15.65	2.5	–	–	2.0	2.0	0.8	–	2	1.6	2	2.0	–	–	–	5
3f	Unscrew 4th of four CRT screws	15.65	2.5	–	–	2.0	2.0	0.8	–	2	1.6	2	2.0	–	–	–	5
3g	Remove CRT/PCB assembly	18.90	–	3	–	4.0	2.5	2.4	1	–	1.6	1	2.5	–	–	–	5
4	Remove CRT																
4a	Remove yoke assembly	13.13	–	3	–	2.0	2.0	1.4	1	–	1.0	1	1.6	–	–	–	1
4b	Remove deflection wire lead	13.02	–	3	–	2.0	2.0	0.8	1	–	1.0	1	1.6	–	–	–	1
4c	Remove CRT	18.06	–	3	–	4.0	2.5	2.2	1	–	1.6	1	2.0	–	–	–	1
5	Remove main wire lead																
5a	Remove 1st retainer screw	15.65	2.5	–	–	2.0	2.0	0.8	–	2	1.6	2	2.0	–	–	–	5

Table 2.6. (continued)

Task number	Task description *Nondestructive disassembly of computer monitor*	Task total	Intersurface friction	Intersurface wedging	Material stiffness	Component size	Component weight	Component symmetry	Force exertion	Torque exertion	Dimensions	Location	Accuracy of tool placement	Posture allowance	Motions allowance	Manpower allowance	Visual fatigue allowance (%)
			(Dis)assembly force			Material handling			Tooling		Accessibility and positioning			Allowances			
8b	Remove swivel support	10.92	–	2	–	2.0	2.0	1.2	1	–	1.0	–	1.2	–	–	–	5
9	Remove rear board																
9a	Bend 1st retaining lug	23.65	–	–	6.5	3.5	2.0	1.4	3	–	1.6	1	2.5	–	–	–	10
9b	Bend 2nd retaining lug	23.65	–	–	6.5	3.5	2.0	1.4	3	–	1.6	1	2.5	–	–	–	10
9c	Bend 3rd retaining lug	23.65	–	–	6.5	3.5	2.0	1.4	3	–	1.6	1	2.5	–	–	–	10
9d	Bend 4th retaining lug	23.65	–	–	6.5	3.5	2.0	1.4	3	–	1.6	1	2.5	–	–	–	10
9e	Remove rear board	10.92	–	2	–	2.0	2.0	1.2	1	–	1.0	–	1.2	–	–	–	5
		698.0	Total score for disassembly operation. Total time for disassembly = 6980 TMUs = 4.188 minutes														

Total maintenance time: 4.188 minutes

Task #1 for disassembly analysis: Remove PCB assembly

Most feasible cost-effective design solution: Replace four retaining lugs with two or use snap-fits to hold PCB assembly in place.

Conclusion: Most amount of time is spent in bending the retaining lugs. Too many lugs hamper disassembly.

Gungor, A., & Gupta, S. (1997). An evaluation methodology for disassembly process. *Computers and Industrial Engineering, 33*(1–2), 329–332.

Gungor, A., & Gupta, S. (1998). Disassembly sequence planning for products with defective parts in product recovery. *Computers and Industrial Engineering, 35*(1–2), 161–164.

Krikke, H. R., van Harten, A., & Schuur, P. C. (1998). On a medium term recovery and disposal strategy for durable assembly products. *International Journal of Production Research, 36*(1), 111–139.

Kuo, T. C. (1999). Disassembly sequence and cost analysis for electromechanical products. *Robotics and Computer Integrated Manufacturing, 16*, 43–54.

Kuo, T. C., Huang, S. H., & Zhang, H. C. (2001). Design for manufacture and design for "X": Concepts, applications, and perspectives. *Computers and Industrial Engineering, 41*, 241–260.

Lambert, A. J. D. (1999). Linear programming in disassembly/clustering sequence generation. *Computers and Industrial Engineering, 36*, 723–738.

Lee, S. G., & Khoo, M. K. (2001). A multi-objective methodology for evaluating product end-of-life options and disassembly. *International Journal of Advanced Manufacturing Technology, 18*, 148–156.

Thierry, M., Salomon, M., van Nunen, J., & van Wassenhowe, L. (1995). Strategic issues in product recovery management. *California Management Review, 37*(2), 114–135.

Zhang, H. C., Kuo, T. C., Lu, H., & Huang, S. H. (1997). Environmentally conscious design and manufacturing: A state-of-the-art survey. *Journal of Manufacturing Systems, 16*(5), 352–371.

Zussman, E., Kriwet, A., & Seliger, G. (1994). Disassembly oriented assessment methodology to support design for recycling. *Annals of the CIRP, 43*(1), 9–14.

CHAPTER 3

ENHANCING DURABILITY OF CONCRETE BRIDGE DECKS BY USING INDUSTRIAL BY-PRODUCTS AS MINERAL ADMIXTURES

Sukhvarsh Jerath
Civil Engineering
University of North Dakota

Charles J. Moretti
Civil Engineering
University of North Dakota

INTRODUCTION

A major cause of deterioration of concrete bridges is the corrosion of steel reinforcement caused by the penetration of moisture and chloride into the concrete mixture. Corrosion reduces the cross-sectional area of the steel, which decreases the stiffness and strength of the structure. The rust formed as the steel corrodes can also cause the concrete to crack due to expansive forces inside the concrete. In many cases the damaging chloride comes from deicing salt used for controlling snow and ice on the bridge deck. Since use of deicing salts is likely to continue, it is important that new bridges be designed to resist chloride-induced corrosion. This can be accomplished by preventing chloride from reaching the steel reinforcement and by making the reinforcement resistant to corrosion. Epoxy-coated steel is widely used in bridge decks to make reinforcement resistant to corrosion, but it does not prevent penetration of moisture and chloride from the surface to the steel reinforcement. Thus, if the protective layer breaks down, the steel is vulnerable to corrosion. Lowering the permeability of the concrete has several beneficial effects. Since chloride usually has to be dissolved in water in order to migrate through the concrete matrix, restricting the movement of water will also help to restrict the movement of chloride. Moisture penetration is also a major cause of cracking in concrete. Cracks form when moisture in the pores of the concrete freezes and expands. Thus, by reducing the permeability of the concrete, moisture intrusion into the pores is restricted and freeze-thaw cracking is reduced.

Corrosion of steel reinforcement (i.e., rebar) in concrete bridges is an electrochemical process, similar to what happens in a battery. Corrosion results from a flow of electrons between anodic and cathodic sites on the rebar. In order for corrosion to proceed, an electron conductor must connect the anode and the cathode. The rebar usually serves as the conductor. There must also be an electrolyte present and in contact with the rebar to complete the electrochemical circuit. The electrolyte is a medium capable of conducting electric current by ionic current flow. Environmental constituents such as chloride and oxygen can establish a suitable electrolyte solution within the concrete matrix when they penetrate the concrete along with moisture (Lambert, 2002).

Reinforcing steel in concrete does not corrode extensively under normal conditions due to the formation of a surface film of iron oxide. The highly alkaline environment (i.e., pH value of about 13) normally found within the concrete matrix stabilizes the oxide film on the surface of the rebar and tends to prevent further corrosion as long as the alkaline environment exists. When the protective iron oxide film remains intact on the surface of the steel, the rebar is said to be in a passive condition. However, there are two commonly occurring situations that promote the corrosion of rebar in concrete. These are carbonation and chloride contamination. Carbonation is a process by which carbon dioxide from the air moves into the concrete matrix and neutralizes the alkalinity. Carbonation can reduce the pH within the concrete to 8 or 9. At this pH, the protective iron oxide film on the rebar is no longer stable. When the protective iron oxide film breaks down, the surface of the steel is exposed to the surrounding concrete matrix. If an adequate supply of oxygen and moisture exists at the exposed surface of the rebar, corrosion will occur. Carbonation is usually a slow process that is controlled by the rate at which carbon dioxide can penetrate into the concrete.

Chloride ions also cause corrosion by disrupting the passive iron oxide film on the reinforcing steel. Two major sources of chloride contamination to concrete are from deicing salts or from seawater in marine environments. Other sources include certain concrete admixtures, contaminated aggregates, contaminated mixing water, airborne salts, and salts in groundwater. The chloride concentration required to initiate corrosion in concrete is relatively low. Field experience and research have shown that on existing structures exposed to chloride ions, a threshold concentration of about 0.026% (by weight of concrete) is sufficient to break down the passive film and initiate corrosion on the reinforcing steel. This chloride concentration equates to 260 parts per million or about 1.0 lb/yd^3 of concrete (Daily, 1995). Another reference indicates that a chloride content of 1.5 lb/yd^3 of concrete would actively support corrosion (Kessler, 2001). The carbonation and chloride penetration can be controlled by reducing the permeability of concrete.

EXPERIMENTAL PROGRAM

Bridge Deck Construction

In this study three bridges were constructed on I-29 in Fargo, North Dakota, United States, in the summer of 2002. It was intended to know whether the permeability of concrete will be reduced by adding mineral admixtures, fly ash, and ground granulated blast furnace slag (GGBFS) as partial replacement of Portland cement. One bridge was constructed with conventional Portland cement concrete, the second bridge was constructed with a modified concrete containing type C fly ash, and the third bridge contained GGBFS of grade 100 as a partial replacement of Portland cement in its concrete. Portland cement was replaced with fly ash and GGBFS by 38% and 35%, respectively, by weight in the two bridges. Corrosion, chloride content, and temperature monitoring activities were carried out at the three bridges between 2002 and 2008. The bridges were monitored to evaluate corrosion of reinforcing steel, chloride content in concrete, and freeze-thaw cycles in the top 5 inches of the bridge deck. Instruments were embedded in the concrete bridge decks during construction to measure corrosion rate and temperature in the bridge decks at various depths in the slab and locations along the bridge span. Additionally, samples were extracted to determine the chloride ion concentration in the concrete bridge decks. For concrete mix designs for the bridges, see Table 3.1, where the coarse aggregate size No.3 refers to the coarse aggregate grading given in the North Dakota Department of Transportation (NDDOT) Standard Specifications for Road and Bridge Construction (NDDOT, 2008).

Monitoring Corrosion in Reinforcement

Corrosion rate in concrete bridge decks was monitored by measuring corrosion rate in steel reinforcement. A patented instrument named Gecor 6 was used to measure corrosion rate in the rebars four times a year, utilizing the polarization resistance technique (NDT James Instruments Inc., 2010). To allow the use of the Gecor 6 corrosion field test instrument, electrical contacts were attached directly to the rebar, close to the points where the corrosion rate measurements were to be taken. To monitor the bridge decks, 15 contacts were attached to each bridge at approximately equidistant intervals along the east edge of the deck. Each contact consisted of a stainless steel rod attached to a small square stainless steel plate. The rod was attached directly to the rebar, and the plate was set level with the deck surface so that it could be accessed for the Gecor 6. When the contact was attached to the rebar, a clamp was used to penetrate the epoxy coating and make electrical contact with the reinforcing steel bar. A picture of a contact attached to the deck rebar is shown in Figure 3.1.

When a Gecor 6 corrosion measurement is taken on a bridge, an electrical lead from the Gecor 6 is attached to the contact plate on the bridge deck and another probe is placed on the concrete over the rebar connected to the contact. All of the contacts were connected to rebar that run transverse to the length of the deck. To take a measurement, the Gecor 6 probe was placed at least 4 inches away from the contact plate on a line with the plate and

Table 3.1. Concrete Mix Design for Bridge Decks

Bridges	*17th Avenue South*	*Texas Turn*	*9th Avenue South*
Cementitious material (lb/cy)	611	611	611
Portland cement (lb/cy)	397	379	611
Fly ash (lb/cy)	0	232	0
GGBFS (lb/cy)	214	0	0
Coarse aggregate size	No.3	No.3	No.3
Max. water/cement (gals/sack)	5.41	5.00	5.00
Max. w/c ratio	0.480	0.443	0.443
Air content (%)	5.0–8.0	5.0–8.0	5.0–8.0
Max. slump (in.)	3	3	3
Design 28-day comp. strength (psi)	4000	4000	4000

1 in. = 25.4 mm; 1 lb/yd^3 = 0.593 kg/m^3; 1 psi = 0.0069 MPa; 1 gal = 3.8 L; 1 sack = 94 lb; 1lb = 0.45 kg

Figure 3.1. Stainless steel contact attachment to the rebar of bridge deck.

perpendicular to the Jersey barrier. The probe was placed on the side of the plate away from the Jersey barrier. When a measurement was taken with the Gecor 6, the diameter of the rebar was used for the corrosion rate calculation. Rebars with a 5/8-inch diameter were used in the 9th Avenue South and 17th Avenue South bridge decks, and 3/4-inch diameter rebars were used for the Texas Turn bridge deck.

The Gecor 6 instrument measures the corrosion rate in the rebar using the linear polarization resistance (LPR) technique. It has been established with laboratory studies that corrosion current is linearly related to polarization resistance. The LPR measurement is done with a central reference electrode surrounded by an external counter electrode. A guard ring and external electrode system confine the area of the rebar tested (NDT James Instruments Inc., 2010). The Gecor 6 gives the corrosion rate (I_{corr}) measurement directly, which is a quantitative measurement of the amount of steel turning into oxide at the time of measurement. I_{corr} is defined by

$$I_{corr} = \frac{B}{R_p} \quad (1)$$

where R_p = polarization resistance, which is defined as the change in potential as measured by the Gecor 6, divided by the applied current; and B = constant equal to 26 mV.

In general, the higher the measured I_{corr} values, the higher the corrosion rate. Gecor 6 measurements should be taken at strategic locations. Corrosion rates will vary depending on conditions, such as concrete moisture content, chloride concentration, and temperature. Thus measurements should be carried out over time in order to obtain average I_{corr} values.

Chloride Content

Measuring chloride content in concrete is important because it indicates the ionic strength of the electrolyte provided by the pore water. Concrete samples were collected once per year from each bridge deck for chloride testing. The samples were obtained by drilling into the concrete decks and collecting drill dust at the depths of approximately 0.25 to 0.75 inches, 2.5 to 3.0 inches, and 4.5 to 5.0 inches. Three samples, one at each depth, were collected at six different locations on each bridge deck. These samples were analyzed for chloride content by using the American Association of State Highway and Transportation Officials (AASHTO) T260-97, Procedure A (AASHTO, 2004, pp. T260-1–15).

Temperature

Temperature has competing effects on corrosion rate; the oxidation of steel increases as the temperature increases and tends to increase corrosion rate. However, because humidity decreases with the increase in temperature, increased temperature can also decrease corrosion rate. Six sets of thermocouples were installed at various points on each of the three bridge decks. Each set consisted of three thermocouples placed at depths of 0.5, 3.0, and 5.0 inches below the surface of concrete. Redundant thermocouples were also installed at each location in case there is damage to any of the thermocouples.

Each Teflon-insulated thermocouple is attached to a wire that runs through a conduit to a data collection apparatus located below the bridge deck. The data logger is an Omega OM-320 microprocessor equipped with an Omega OM-320-HLIM-1 analog interface module and is capable of storing more than 32,000 data points. Each data logger is housed in a steel box attached to a bridge pier, and a heater is attached in the storage box to protect the equipment from cold temperatures. The temperature-collecting apparatus is programmed to collect more data as the temperature approaches freezing and the rate of change of temperature increases. Temperatures were measured to estimate the number of freeze-thaw cycles occurring. The data collection systems are programmed to take temperature readings in degree Fahrenheit from the 18 points where the thermocouples are inserted in each bridge deck.

The positions of thermocouples A1, A4, B1, B4, C1, and C4 are indicated in Figure 3.2. These six thermocouples are positioned 0.5 inches below the surface of the concrete deck. Thermocouples A2 and A3 are positioned directly below A1, 3 inches and 5 inches below the surface, respectively. Thermocouples A5 and A6 are positioned below A4, 3 inches and 5 inches below the surface, respectively. Similar designations were used for the thermocouples sets designated B and C. Thus there are 18 thermocouples for temperature readings in each bridge.

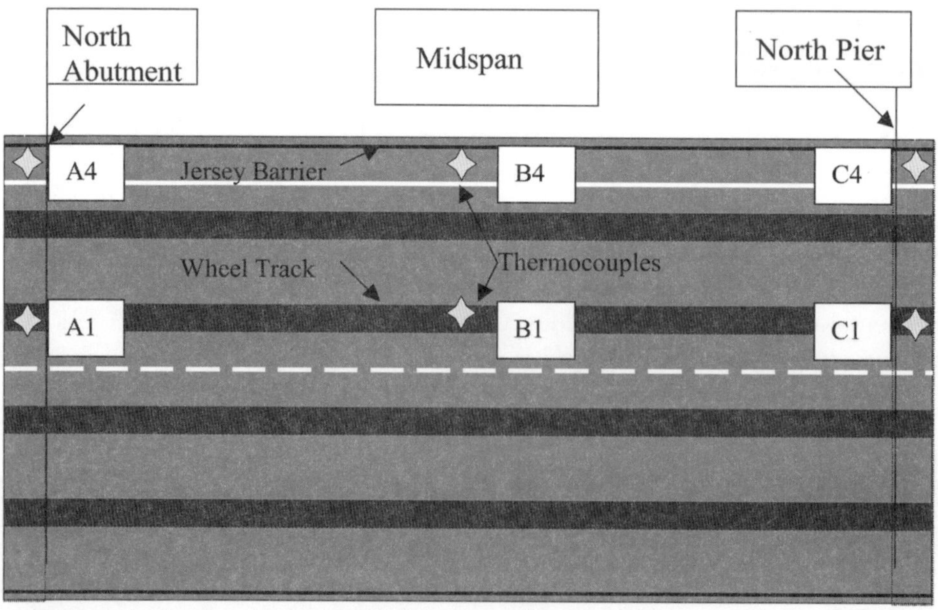

Figure 3.2. Location of thermocouples on bridge deck.

TEST RESULTS

I_{corr} data obtained with the Gecor 6 meter are a quantitative measure of the amount of steel turning into oxide at the time of measurement. The following broad criteria have been established to interpret corrosion information provided by the Gecor 6 meter:

- I_{corr} less than 0.2 μA/cm^2 = passive condition;
- I_{corr} between 0.2 and 0.5 μA/cm^2 = low corrosion rate;
- I_{corr} between 0.5 and 1.0 μA/cm^2 = moderate corrosion rate; and
- I_{corr} greater than 1.0 μA/cm^2 = high corrosion rate.

Measurements were taken at 15 almost equally spaced locations on each bridge deck. The locations are numbered 1 through 15, with 1 situated at the north end of the deck and the increasing numbers extending sequentially along the span to the north pier. The locations where the measurements were taken are close to the Jersey barrier on the east side of the deck. The sampling plan specified that four I_{corr} measurements be taken at each location between August and June each year. The average I_{corr} measurements taken on the bridge decks over the 5 years for each location are given in the tabular form in the report (Moretti, 2008). Air temperature and relative humidity were also recorded on the bridges when the measurements of I_{corr} were taken. The temperature and humidity data corresponding to the various sampling dates are also listed (Moretti, 2008).

The overall corrosion rates for the rebar in the bridge decks were estimated by averaging the I_{corr} values collected from all locations on the deck for each sampling date. The overall corrosion rates are plotted in Figure 3.3. The trend lines on Figure 3.3 were generated by linear regression on the overall data from each bridge. These lines indicate how corrosion is progressing with time. The trend lines suggest that the corrosion rate is slowly increasing with time. This is expected because environmental factors that support corrosion, such as moisture, oxygen, and chloride, tend to diffuse deeper into the concrete over time. All of the I_{corr} values in Figure 3.3 are below 0.2 μA/cm^2. These results indicate that the rebar at all measurement locations is in a passive (i.e., noncorroding) state. This is reasonable because the decks were monitored within 5 years of their construction and the rebars have epoxy coating to protect them from corrosion.

$$M = \frac{I\,t\,A_w}{n\,F} \qquad (2)$$

where M = mass of metal dissolved or converted to oxide (grams); I = current (Amps); t = time (seconds); A_w = atomic weight; n = valency; and F = Faraday's constant (96,500 coulombs/equivalent mass).

For iron in reinforcing steel, a factor of 0.492 can be used to convert I_{corr} values to thickness loss in units of mils/yr. (American Concrete Institute, 2001). Rates of steel loss for the rebar were calculated based on overall I_{corr} values averaged over each year of the project. These values are compared in Figure 3.4.

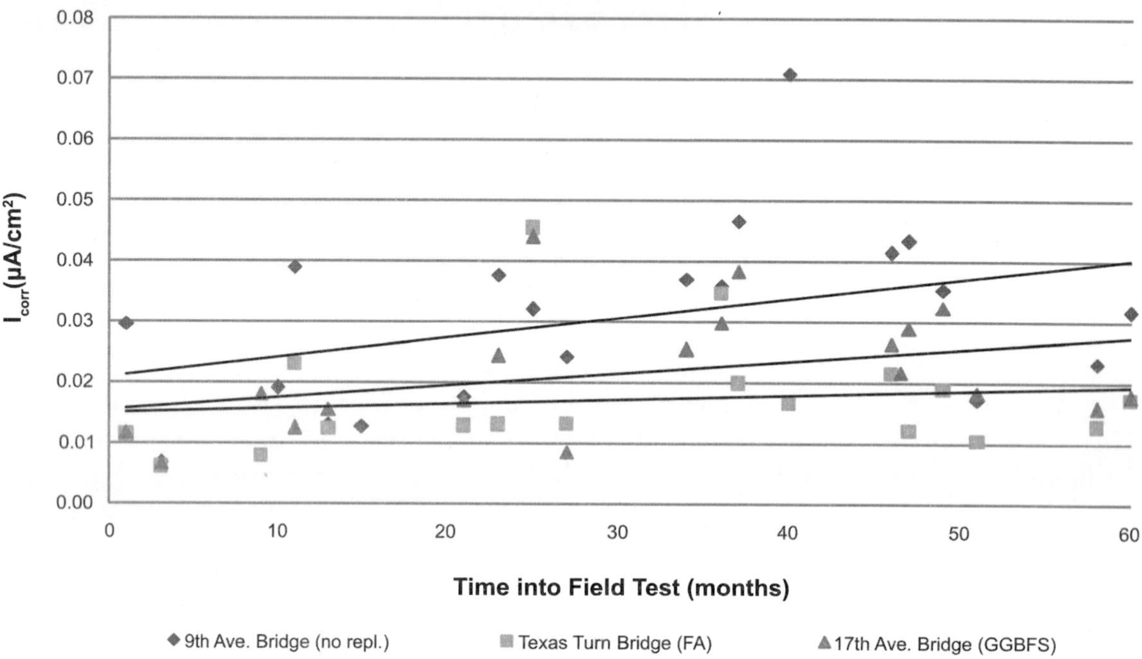

Figure 3.3. Average I_{corr} values on the bridges at different dates and regression lines. An I_{corr} value can be converted to the rate of metal thickness loss using Faraday's law.

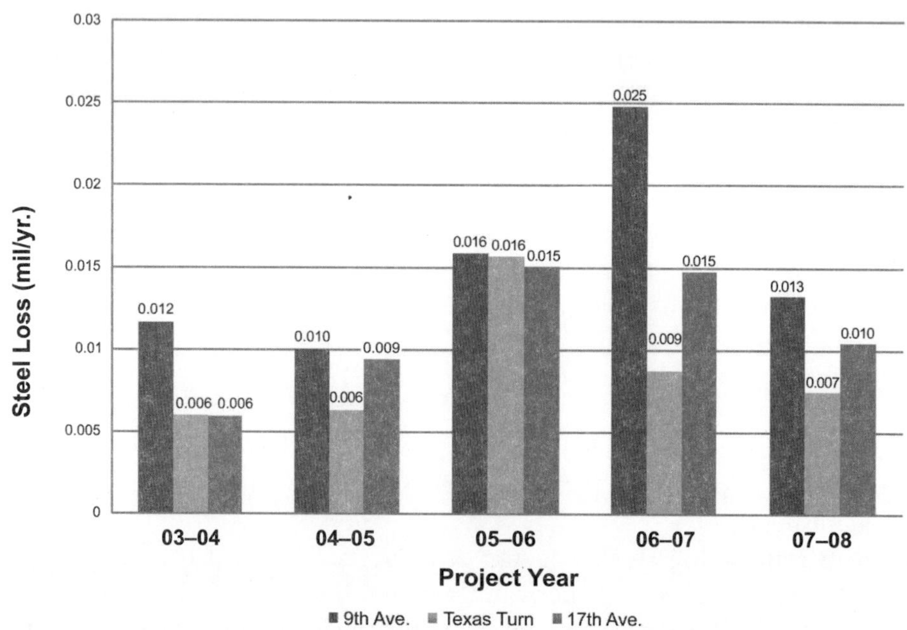

Figure 3.4. Estimated rate of steel loss in rebars during different years.

The results of all the chloride analyses performed on concrete samples obtained from the three bridge decks are listed in the report to the NDDOT (Moretti, 2008). Average annual chloride contents were obtained from the samples taken at three different depths from six locations on each bridge and are seen in Tables 3.2, 3.3, and 3.4.

It is seen that the chloride levels in samples taken at all depths are increasing over time. This is expected since the chlorides are applied to the bridges each winter for deicing. The average chloride concentrations measured in the most recently collected concrete samples on August 2007 are shown in Figure 3.5. The figure shows that the highest average chloride levels were measured for the 9th Avenue Bridge.

Collection of temperature data from thermocouples cast in the three bridge decks was started during the summer of 2002.The major reason for collecting temperature data is to determine how many freeze-thaw cycles the concrete is subjected to over time.

Table 3.2. Average Chloride Concentrations (lb/yd^3) of Samples From 9th Avenue Bridge Deck

Sample depth	*October 2002*	*August 2003*	*August 2004*	*August 2005*	*August 2006*	*August 2007*
0.25–0.75 in.	0.56	0.88	2.80	5.26	9.09	9.84
2.5–3.0 in.	0.57	0.60	0.24	1.03	2.27	3.91
4.5–5.0 in.	0.62	0.64	0.24	1.19	1.33	1.46

1 in. = 25.4 mm; 1 lb/yd^3 = 0.593 kg/m^3.

Table 3.3. Average Chloride Concentrations (lb/yd^3) of Samples From Texas Turn Bridge Deck

Sample depth	*October 2002*	*August 2003*	*August 2004*	*August 2005*	*August 2006*	*August 2007*
0.25–0.75 in.	0.55	0.59	1.92	3.37	3.21	5.70
2.5–3.0 in.	0.65	0.59	0.31	1.15	1.77	1.13
4.5–5.0 in.	0.71	0.69	0.30	0.87	1.73	0.98

Table 3.4. Average Chloride Concentrations (lb/yd^3) of Samples From 17th Avenue Bridge Deck

Sample depth	*October 2002*	*August 2003*	*August 2004*	*August 2005*	*August 2006*	*August 2007*
0.25–0.75 in.	0.69	0.71	1.69	1.86	3.59	3.21
2.5–3.0 in.	0.62	0.61	0.60	1.04	2.38	1.92
4.5–5.0 in.	0.57	0.62	0.28	1.11	1.43	1.44

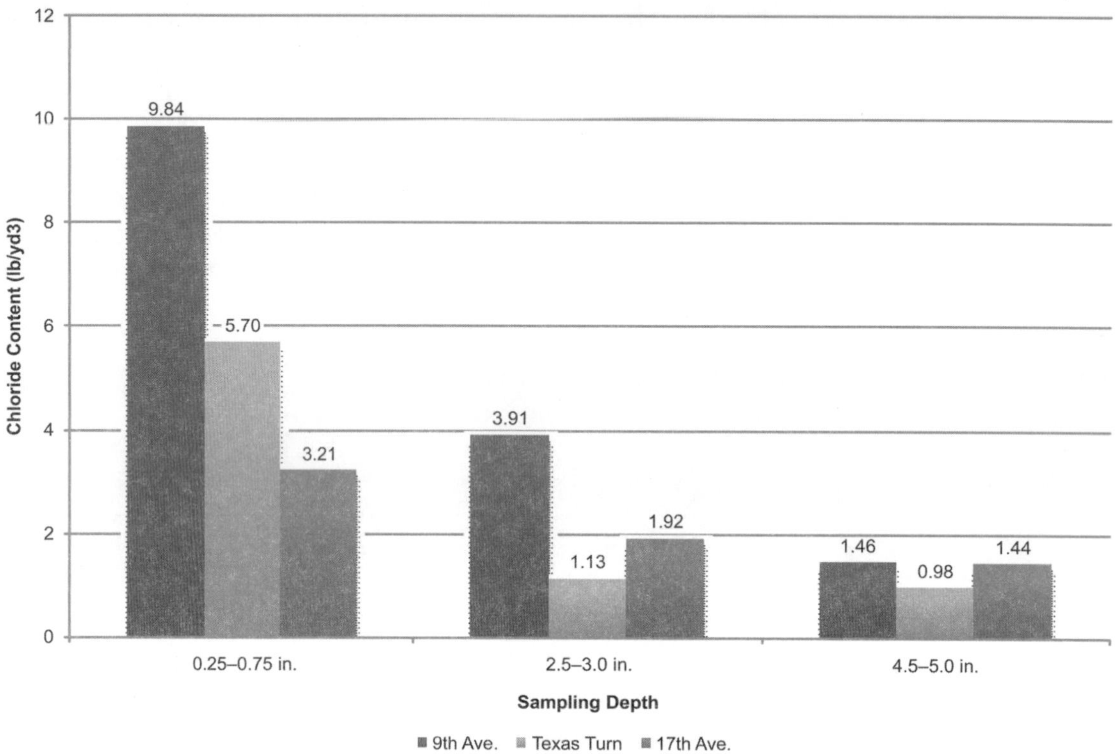

Figure 3.5. Chloride levels at three depths in the bridge decks measured in August 2007.

Repeated freezing and thawing is a concern because it can cause cracks to form in the concrete. Cracking can accelerate corrosion of the steel rebar if it permits entry of moisture, air, and chloride into the concrete matrix. The number of freeze-thaw cycles observed for the three bridge decks is listed in the report to the NDDOT (Moretti, 2008). The maximum and minimum concrete temperatures recorded for each deck at the thermocouple locations B1, B2, and B3 are listed in Table 3.5. These particular probes are located at depths of 0.5, 3, and 5 inches below the surface of the deck at approximately midspan in the left lane of the bridge as shown in Figure 3.2.

The average numbers of freeze-thaw cycles that occurred at each bridge deck are calculated separately at a specific depth and location. The numbers are averaged separately for the probes located in the left traffic lane, and probes located close to the Jersey barrier shown in Figure 3.2. The freeze-thaw numbers are compared for the three decks at specific depths and location for different years in Figures 3.6, 3.7, and 3.8 for the left lane. Figures for the Jersey barrier location are given in the report to the NDDOT (Moretti, 2008).

It is seen from Figures 3.6, 3.7, and 3.8 that the number of freeze-thaw cycles varied significantly from year to year. However, looking at these figures, there does not appear to

Table 3.5. Maximum and Minimum Temperatures

Bridge deck	*Probes*	*Temperature °F*				
		July 2003 to June 2004	July 2004 to June 2005	July 2005 to June 2006	July 2006 to June 2007	July 2007 to June 2008
9th Avenue	B1	118.7/–18.2	119.4/–14.0	121.6/–12.3	127.2/–13.0	124.7/–16.4
	B2	108.5/–16.0	110.5/–11.2	12.8/–9.7	115.2/–10.8	115.4/–14.4
	B3	101.2/–14.7	105.7/–9.5	107.8/–8.8	107.5/–9.5	110.8/–13.3
Texas Turn	B1	115.6/–15.7	116.6/–9.7	118.1/–9.1	125.0/–10.1	121.7/–10.8
	B2	106.9/–13.4	107.5/–5.8	108.2/–6.0	111.0/–7.6	111.5/–7.9
	B3	103.2/–11.6	103.4/–2.9	103.5/–4.0	103.0/–5.4	106.6/–5.8
17th Avenue	B1	111.7/–16.5	113.3/–12.9	113.8/–10.2	122.0/–11.0	119.4/–14.1
	B2	105.6/–14.6	106.4/–10.8	107.2/–8.1	111.4/–9.1	111.4/–12.4
	B3	103.7/–12.7	103.2/–8.9	103.8/–6.4	106.5/–7.2	108.3/–10.3

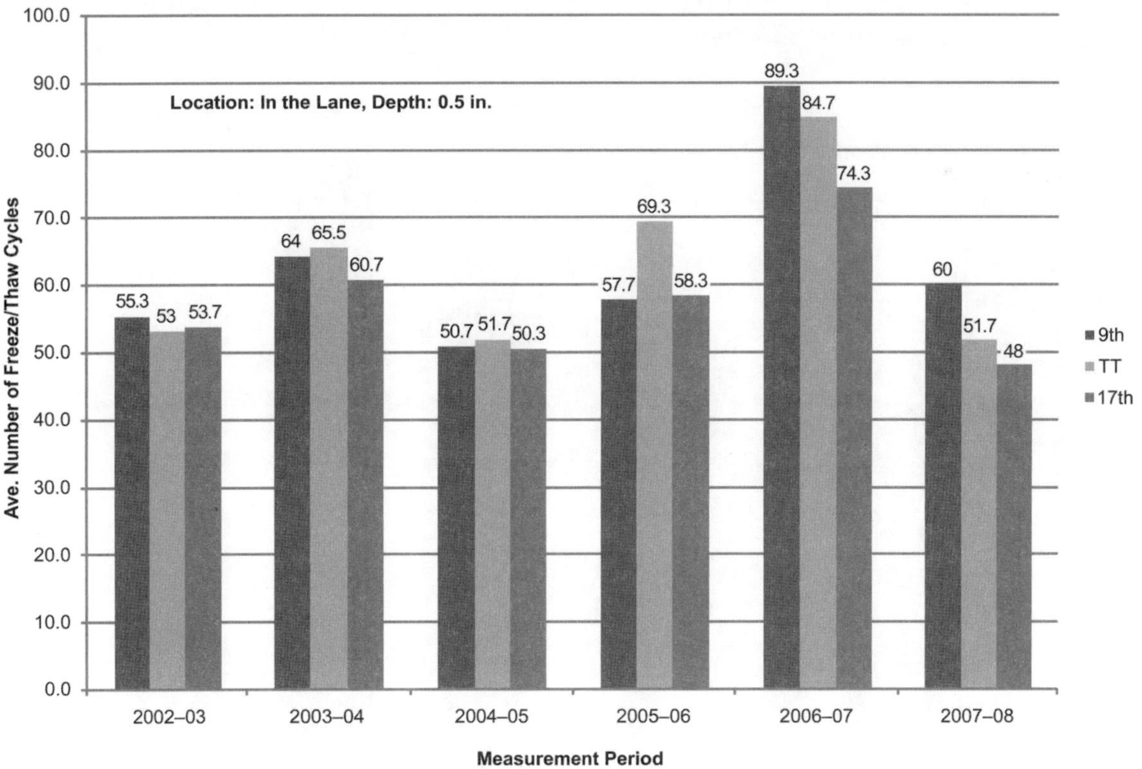

Figure 3.6. Average number of freeze-thaw cycles in the bridge decks in the left lane at a depth of 0.5 inches.

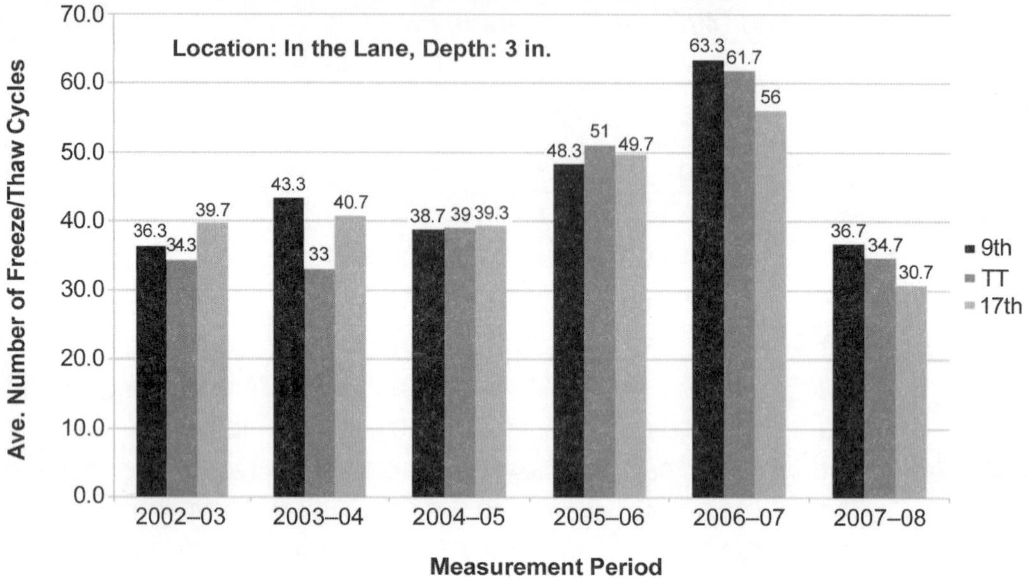

Figure 3.7. Average number of freeze-thaw cycles in the bridge decks in the left lane at a depth of 3 inches.

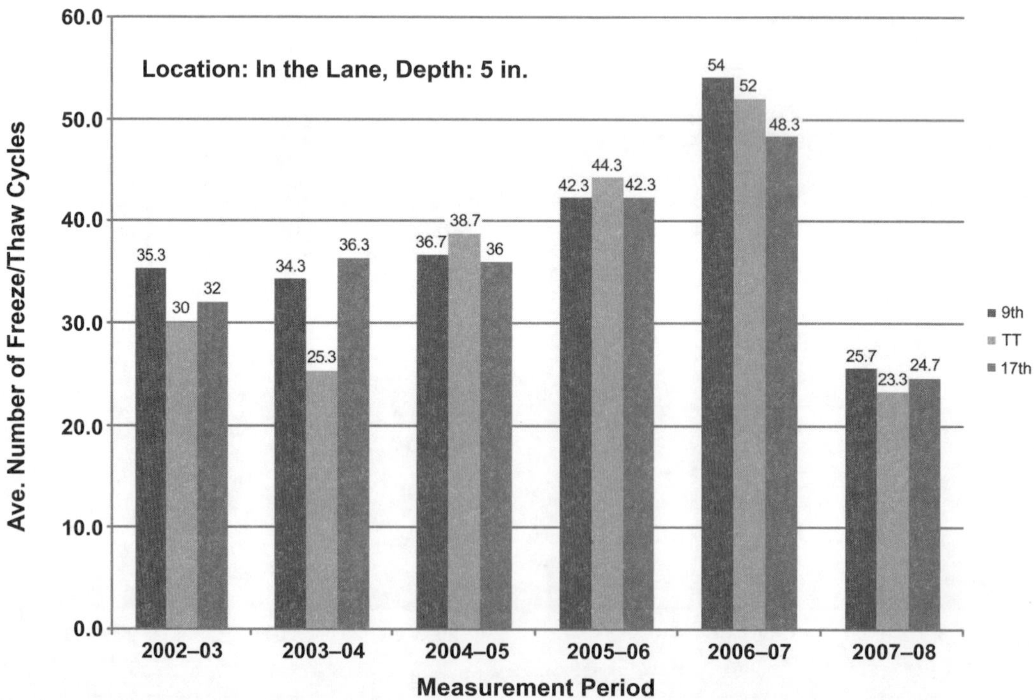

Figure 3.8. Average number of freeze-thaw cycles in the bridge decks in the left lane at a depth of 5 inches.

be a clear trend to suggest that any one bridge experienced significantly more cycles than the others.

Each summer a visual examination of the bridge decks was made to note any obvious deterioration that may have occurred during the previous year. Hairline cracks were observed in all the bridge decks, but none of the bridges had any cracks that represented any significant structural damage.

CONCLUSIONS

1. All three bridges were measured to have I_{corr} values lower than 0.2 $\mu A/cm^2$, indicating that all the rebars were in the passive (non-corroding) condition as of July 2008, 6 years after their construction at the locations tested on the three bridge decks. The bridge containing no mineral admixture produced the highest I_{corr} readings, the bridge containing GGBFS grade 100 had the next high value, and the bridge containing class C fly ash exhibited the lowest corrosion rate. Even though the trend lines in Figure 3.3 indicate a gradual increase in I_{corr} values over the first 4 years of the project, there is a trend that the I_{corr} values begin to decrease in the 5th year. This reduction may be due to less permeability of concrete with time because it becomes denser. It is recommended to continue the monitoring in future years to know how effective are the mineral admixtures in preventing the corrosion of reinforcing bars in the bridge decks.
2. Chloride levels are increasing over time in all the bridges as shown in Tables 3.2, 3.3, and 3.4. These increases occurred at all the depths of 0.5, 3, and 5 inches below the bridge deck at which the chloride contents were measured. Highest chloride levels were found as expected at 0.5 inches below the bridge deck surface, the depth closest to the surface. The highest chloride contents were measured for the 9th Avenue Bridge that did not contain mineral admixture. The lowest chloride contents at 3- and 5-inch depths were measured in the Texas Turn Bridge that contained fly ash mineral admixture. The chloride levels lie in between for the 17th Avenue Bridge that contained GGBFS as partial replacement for Portland cement. The Texas Turn Bridge was the only one of the three bridges where the average chloride content was below 1.0 lb/yd^3 at a depth of 5 inches in the latest reading taken on August 2007.
3. The number of freeze-thaw cycles recorded with the thermocouples is determined by reading through the data files and counting number of times the temperatures went below 32 °F (0 °C) and then above the freezing temperature. The number of these cycles observed for the three bridges is shown in Figures 3.3, 3.4, and 3.5. More freeze-thaw cycles occurred in the traffic lane than the areas closer to the Jersey barrier (Moretti, 2008). Also, more freeze-thaw cycles occurred closer to the surface of the concrete deck (0.5 inches below the surface) than at points deeper in the deck (3 and 5 inches below the surface). The results show that the number of freeze-thaw cycles in the decks varied from year to year. However, comparing results in the same year, there is no clear trend to indicate that any one of the bridges consistently experienced more freeze-thaw cycles than the others. So it can be concluded that freeze-thaw effects were not a major variable in terms of the overall performance of the three bridge decks.

4. One of the key questions this study sought to answer was whether the addition of fly ash or GGBFS to the concrete reduced the rate of intrusion of moisture and chlorides. Based on the results obtained, it appears concrete containing the mineral admixtures performed better than the plain concrete. The chloride data indicates that after the first year of the project, the plain concrete consistently contained the highest chloride levels at 0.5, 3, and 5 inches of depth into the deck. Overall, the bridge decks containing mineral admixtures performed better in reducing permeability, and it seems fly ash is a better mineral admixture.

REFERENCES

American Association of State Highway and Transportation Officials (AASHTO). (2004). *Standard specifications for transportation materials and methods of sampling and testing, part 2B: Tests*. Washington, DC: AASHTO.

American Concrete Institute. (2001). *222R-01: Protection of metals in concrete against corrosion*. Farmington Hills, MI: American Concrete Institute.

Daily, S. F. (1995). *Understanding corrosion and cathodic protection of reinforced concrete structures*. Medina, OH: Corrpro Company.

Kessler, R. (2001). *Corrosion rate based on polarization resistance*. Washington, DC: AASHTO.

Lambert, P. (2002). *Corrosion protection association monograph*. Retrieved from http://www.azom.com/details.asp?articleID=1318

Moretti, C. J. (2008). *Bridge monitoring project, final report*. Bismarck, ND: North Dakota Department of Transportation.

NDT James Instruments Inc. (2010). Rebar corrosion and its effects. Retrieved from http://www.ndtjames.com

North Dakota Department of Transportation (NDDOT). (2008). *Standard Specifications for Road and Bridge Construction, Vol. 1*. Bismarck, ND: NDDOT.

PART III

EQUIPMENT AND MAINTENANCE

CHAPTER 4

RESOURCE UTILIZATION ENHANCEMENT AND OPTIMIZATION THROUGH DESIGNING EQUIPMENT FOR EASE OF MAINTENANCE

Anoop Desai

Department of Mechanical Engineering and Electrical Engineering Technology
Georgia Southern University

Anil Mital

Department of Mechanical Engineering
University of Cincinnati

INTRODUCTION

Human civilization has progressed by leaps and bounds since the beginning of the 19th century. This has been made possible by the development of numerous new products and inventions that have made life simpler and enhanced productivity. One has only to look at the human population, which has witnessed an exponential rise since the 1800s. How did this metamorphosis come about? What really happened in the 19th century that led to this tremendous burst of human creativity? The answer can be found in two simple words: Industrial Revolution.

The Industrial Revolution happened in Europe in the late 18th and early 19th centuries. The discovery of steam power and the invention of the steam engine by James Watt enabled humans to utilize mechanical power to serve their needs. This change had massive repercussions. The age of the machines had dawned, and it continues unabated today, albeit morphed by the many different technologies that have evolved in its wake.

Industrial Equipment and Machines as Important Resources

Machines characteristically are comprised of moving parts (especially those machines that use mechanical power) and use friction to perform work. Friction leads to wear and tear of mating surfaces. The phenomena of wear and fatigue (wear that occurs over an extended period of time) eventually leads to equipment failure.

Machine failure can have catastrophic consequences, especially in light of resources availability. Most machines are made of steel and have other alloys present in them, such as copper alloys, depending on their particular application. It has been established that the world's iron supply or, for that matter, the world's copper supply is dwindling rapidly and cannot keep pace with demand. This is especially true as two of the world's most populous countries, China and India, wake up to industrialization. One only needs to look at copper prices to understand the gravity of the situation. Copper prices have soared from about $0.73/lb in 1999 to roughly $3.00/lb in 2006. Copper prices are often used by economists as a gauge to reflect world economic activity.

Basic economic law dictates that if demand outstrips supply, prices will rise, and if this happens on a large scale, prices will rise exponentially. A look at the Commodity Research Bureau's (CRB) handbooks will reveal that over the course of the past 40 years, very little investment has been made in basic infrastructure that would enable mining of industrial metals as well as precious metals. If new mines are not developed and if the related infrastructure is not updated, how can we expect supply to catch up with demand? Gold, for instance, is a case in point. Gold is a precious metal that is used as a substitute for money. But it is used in numerous industrial settings as well. So is silver. Gold prices have rocketed from about $250 per troy ounce to roughly $1,000 per troy ounce over the course of the past 9 years (2000–2009).

In view of this situation, it is apparent that the world is running out of resources that have industrial applications. This includes both base metals (copper, tin, iron, etc.) as well as precious metals (gold, silver, platinum, etc.) that have wide-ranging industrial applications owing to their excellent physical properties (such as electrical conductivity, malleability, ductility, etc.). It is obviously necessary to conserve these resources for future use and future generations.

Optimizing Use of Industrial Equipment Resources Through Regular Maintenance

One of the means by which these resources can be conserved is through regular maintenance of industrial equipment. In doing so, it will not be necessary to replace equipment so often. Within the constraint of technological currency, it would be possible to optimize the use of current industrial equipment by resorting to regular maintenance.

According to the department of defense, maintainability is "a characteristic of design and installation which is expressed as the probability that an item will conform to specified conditions within a given period of time when maintenance action is performed in accordance with prescribed procedures and resources" (Harring & Greenman, 1965). Designing equipment for maintenance has been practiced more as an art than as a science. It has evolved to a greater extent as a result of common sense than by means of scientific investigation. Maintenance is perhaps the most expensive aspect of industrial equipment use. The following reasons are responsible for this fact:

- Given the proliferation of different types of products and systems, there has been an ever-increasing need to perform different kinds of maintenance operations.
- The cost of human labor has increased consistently. An estimation of the cost of human labor is extremely important since maintenance is perhaps the only field of operation that relies solely on human capital and human skill.

This chapter addresses the design for maintenance practice. It is divided into two parts. The first part examines various maintenance concepts that are applied to regular preventive maintenance. The second part examines various designs for maintenance algorithms and related practices. It also presents a proactive design for maintenance algorithm developed by the authors and its practical application.

MAINTENANCE CONCEPTS

The rapidly evolving complexity of machines has kept pace with evolving technology. Improvements in reliability techniques, however, have been unable to keep pace with the growing degree of machine complexity. New problems in equipment downtime have been proliferating, and the concept of maintenance to serve as a tool to reduce downtime has assumed a growing importance (Imrhan, 1991). This section will describe some of the basic concepts related to maintenance and related procedures.

As far as designing equipment for maintenance is concerned, it has been practiced more as an art than as a science. This is due to the following reasons:

- The increasing need to perform maintenance activities.
- High and increasing cost of human labor. An estimation of the cost of human labor is extremely important since maintenance is perhaps the only field of operation that relies solely on human capital and human skill.

Some examples from the military aircraft industry that corroborate this claim are presented as follows:

- Aircraft maintenance costs in the United States have been estimated to amount to approximately 35% of life cycle costs of military systems (McDaniel & Askrein, 1985).
- The technical complexity of modern aircraft has compounded the problem of quick, cost-effective maintenance even more. Adding to the degree of complexity is the wide array of hardware made possible by computer-aided design systems.
- In 1970, the U.S. Department of Defense allocated a quarter of its budget to maintenance costs. This fraction is expected to grow with increasing level of complexity of components and machine assemblies.

From this discussion, it is clear that machines and products that are designed with a view to enhance the ease of maintenance lend themselves more easily to that particular function. This results in maintenance operations being performed at a fraction of their regular costs and at a fraction of the time required otherwise. In this context, the importance of designing for maintenance cannot be overemphasized. However, very little research work has really been performed with a view to enhance maintainability of products and machines. A distinction needs to be drawn between functional design (design for operability, in this context) and design for maintenance.

Designing for maintenance is more difficult than designing for operability due to the following reasons:

- Environmentally speaking, the maintenance workplace is much more variable as well as unpredictable. Maintenance technicians are often forced to work in limited and cramped workspaces, which in turn have not been designed to perform maintenance operations.
- The degree of variability inherent in equipment is staggering. This is truer in the case of consumer products such as consumer electronics, for instance. This problem is further compounded by the rapid degree with which equipment becomes obsolete. This, in turn, underlines another problem: the training and retraining of the maintenance crew. However, this point is not within the scope of this chapter, hence it is not appropriate to discuss it here.
- A crucial point of difference is the obviousness of conflicting goals as far as the maintenance and operation of equipment is concerned. For instance, clearance within a machine and between a machine's parts is crucial to enable maintenance. However, it may not always cater to enhance operability of the product. The present trend is toward miniaturization. This trend is driven by the need to lower cost of production, to ease the manipulation of machines during operation and transportation, and to satisfy the demands of a gradually shrinking workplace site. The issue of designing a workplace site, while relevant to the general scope of maintenance, is not directly related to machine design for maintenance. Hence, it will not be alluded to in detail in this chapter.

Maintenance-friendly machines are important, in terms of both production as well as safety. Also, machines that are difficult to maintain routinely are less likely to receive the required standard of maintenance. For example, the breakdown of some machines was often found to be associated with, or was a direct result of, the lack of maintenance or abuse of equipment rather than just poor engineering. It is clear that although there have recently been significant improvements in reliability, it will, in all likelihood, not be affordable to develop a totally reliable machine. For this reason, good maintainability will always be important.

There are four principal methods by which maintenance can be acheived:

Corrective maintenance. This is achieved by adopting the following steps:

- Failure detection leads to confirmation. If the failure is not confirmed, the item must be returned to service. This no-fault-found problem leads to a considerable amount of time being wasted at significant cost. It also entails the carrying of an unnecessarily large amount of inventory all the time.
- If the failure is confirmed, the item is prepared for maintenance action and the failure report is completed.
- Localization and isolation of a failed part in the assembly is the natural next step in performing corrective maintenance.
- The failed part is removed for disposal or repair. If it is disposed, a new part is installed in its place. Examples of repairable parts or connections include broken connections, open circuit board on a printed circuit board, or a poor solder.
- The item may be reassembled, realigned, and adjusted after repair. It is checked before being put back to use.

Predictive maintenance. Predictive maintenance is also referred to as preventive maintenance. Preventive maintenance is carried out so as to minimize the probability of a failure. It is often referred to as use-based maintenance (Swanson, 2001). This type of maintenance is comprised of maintenance activities that are undertaken after a specific amount of time or after a specific amount of equipment use (Gits, 1992; Herbaty, 1990). It relies on the estimated probability of equipment failure in the given interval of time.

Maintenance of a degrading system. Most systems operate with some sort of degradation occurring throughout their useful lives. In order to enable the maintenance of such systems, a review has to be done periodically to determine what maintenance actions need to be taken. In order to optimize the maintenance schedule, it has been suggested that the level of degradation be monitored instead of time. This approach enables the addition of factors such as maintenance costs and distribution of degradation. After each monitoring period, the amount of degradation is measured. Maintenance is carried out if degradation exceeds the amount specified. The amount of degradation is assumed to be a nonnegative, continuously random variable, and for each monitoring period it is the same and independently distributed. It is to be understood that an optimal maintenance plan obviously depends on cost factors. As such the costs of overhauling and operating a system must be included in the evaluation. It seems that this approach reduces long-term discounted costs.

Aggressive maintenance. It is clear from its nomenclature that aggressive maintenance implies a maintenance philosophy that is much more aggressive and far-seeking than preventive maintenance. An aggressive maintenance strategy seeks to improve overall equipment operation. Aggressive maintenance draws upon the concept of total productive maintenance (TPM). Hence, it is essential to understand the concept of TPM in order to fully realize the benefits of aggressive maintenance.

TPM may be defined as a partnership approach to maintenance. It is a philosophy that chiefly deals with maintenance management designed in order to complement the implementation of just-in-time systems in Japanese plants (Swanson, 2001). In its basic form, a just-in-time system seeks to replace inventory as existing inventory gets depleted. This system allows customers to drive the inventory replenishment process thus reducing inventory carrying costs. This practice leads in a lean organization. TPM activities seek to eliminate the "six major losses" related to equipment maintenance. These losses are comprised of equipment failure, setup and adjustment time, idling and minor stoppages, reduced speed, defects in process, and reduced yield. Under TPM, small groups or teams create a cooperative relationship between maintenance and production that ultimately aids in the accomplishment of maintenance tasks. Also, given the team nature of work, production workers get involved in performing maintenance work, thereby allowing them a role in equipment monitoring and upkeep. This consequently raises the skill of production workers and raises their efficiency in maintaining equipment.

Maintenance prevention teams work to improve equipment performance through improved equipment design (Swanson, 2001). To this end, the maintenance department works cohesively with the engineering department during the early stages of design. The result is equipment that is easy to operate and maintain.

The chief advantage of TPM (and hence aggressive maintenance) is the obvious improvement in equipment availability and reduction in maintenance costs. This further leads to better maintenance efficiency and reduced repair time.

DIFFERENT DESIGN METHODOLOGIES FOR MAINTENANCE

There are numerous mathematical models that seek to address the problem of effective maintenance through objective, numerical problem formulation. However, there is a basic challenge as far as this approach is concerned. The success of this approach, which is strictly a branch of applied mathematics or statistics, can only be measured in terms of its impact upon the solution of real maintenance problems. It should be pointed out that a major problem exists with this approach. While new theories keep appearing at an unprecedented rate, too little attention is paid to data collection as well as to the consideration of the usefulness of models for solving real problems through model fitting and validation.

When speaking in terms of mathematical models, simple models are those that contain a small number of unknown parameters. An example of a simple model would be that proposed by Barlow and Hunter (1960). It proposed an age-based replacement model for a component. This component is renewed on replacement using a two-parameter, Weibull time-to-failure distribution. In this model, only a small number of observations of the time before the equipment fails (approximately 10) would be required. This would enable determination of the optimal or near optimal value of the critical age at which preventive maintenance should be carried out. One of the chief disadvantages with this model is the obvious scarcity of real-life examples that require a data set of such miniscule dimensions. As a result, the practical validity of this model is highly suspect.

There are other models that are more complex in their problem formulation (with respect to number of unknown parameters). One of the chief drawbacks of these models is that the degree of correlation between the different parameter estimates is high. In other words, the parameter estimates tend to overlap each other to a large extent. This means that the proposed model is unable to distinguish clearly between different parameter combinations. This leads one to be quite certain about the inefficacy and invalidity of problem formulation in the first place. Complex models often do not have the necessary and sufficient data required to arrive at a solution. Another drawback with these models is that they present a very complicated solution and are unable to make accurate (or even feasible) predictions.

Finally, it has to be mentioned that while mathematicians and statisticians tend to be intent on enhancing the complexity of their models, management and engineering are looking for simple, straightforward, and transparent models to solve what is essentially a very practical problem. Theoretical solutions arrived at by solving highly complicated mathematical equations (often with very little relevant data available to solve them) offer very little by way of a real solution to a real problem.

Capital replacement modeling is an area of maintenance that is considered by some as strategic or long-term maintenance. It is considered to be a part of strategic planning. Strategic planning consists of the provision of resources in order to safeguard an organization's future competitiveness. It has been pointed out that while technological and economic factors may be the principal drivers for equipment replacement, maintenance costs and unavailability are just as important. From the ongoing discussion, it is clear that mathematicians have had the most influence on problem formulation and the most attempts to solve them.

The development of condition monitoring coupled with that of decision models has put new demands on maintenance management information systems (MMIS). Currently, there are a substantial number of systems available with the goal of managing maintenance. The specific objectives of such systems are as follows:

- Track specific components through the maintenance cycle.
- Provide logistic support to plant managers, maintenance engineers, and maintenance personnel. Example of logistic support include providing information on and tracking spares inventory.

- Record, maintain, and provide store maintenance history.
- Alarm predetermined maintenance activity.
- Produce management reports to enable strategic actions (such as aggressive maintenance).

While a large majority of MMIS are able to accomplish the previously mentioned objective satisfactorily, a small number of systems are also able to perform the following auxiliary functions:

- Analyze maintenance history.
- Determine optimal policy for components and subsystems. This is tantamount to optimizing inventory control of spare parts inventory, thereby helping reduce spare parts overhead costs substantially.

In case of large and complex systems with many subsystems and component interactions, MMIS will have to provide solutions related to the following areas:

- Incorporate expert opinion in a knowledge base.
- Include subjective data from experts in the maintenance field and related fields.
- Draw up a schedule of maintenance activities (especially from the perspective of preventive and aggressive maintenance).
- Update maintenance schedules with the occurrence of operational events such as component or system failures as well as unscheduled replacements.
- Plan resources.
- Measure the effectiveness of maintenance activities. It should be noted that the formulation of an objective index would be most helpful in this regard.

Mathematical modeling yields solutions to strictly theoretical problems. They are not effective in tackling problems commonly occurring in industry. Similarly, all mathematical modeling is essentially reactive in nature. It attempts to solve a problem after it has occurred. There is no attempt to prevent a problem from occurring to begin with. The proactive approach to maintenance is a design issue. The next section of this chapter provides an overview of the different design approaches used by researchers to attempt to design products or systems for ease of maintenance.

PREDICTION MODELS FOR MAINTENANCE

Prediction procedures for maintainability enable the designer to forecast the effects of design on system repair. The findings of maintainability prediction indicate the extent to which design contributes to ease of support. As a result, it is easy to estimate what additional maintainability features will be required (Harring & Greenman, 1965). Prediction models indicate the downtime to be expected from a system prior to its operation in real-life situations. It also points to which of the system's features are likely to cause serious trouble. Maintainability predictions complement qualitative design parameters. This is important since the maintenance engineer is often concerned with system availability and must resort to quantitative criteria in order to measure the effects of qualitative design features.

The Radio Corporation of America (RCA) Method

The RCA method utilizes support time as the criterion of maintainability. It is essentially a technique that uses a checklist of physical features of product design. Support time is regarded as a function of the following attributes:

- Physical design features
- Support requirements
- Personnel requirements essential in order to affect efficient maintenance

Design features play a pivotal role in evaluating the physical aspects of a system. They are also utilized to determine the effects of layout, accessibility, and packaging on support time. Physical design of a product is evaluated on the basis of 15 sets of questions. Each question is assigned a value in terms of the impact of physical design on repair time.

A linear equation is developed for support time by the regression analysis of the empirical data (produced by more than 100 support incidents occurring in the operation of ground electronic equipment). This is presented as follows:

$$Z = 3.54651 - 0.02512A - 0.03055B - 0.01093C$$

A, B, and C represent measures of the three parameters.

As far as the three elements affecting support time are concerned (as mentioned earlier), the group of features listed is regarded as the most significant.

The Federal Electric Method

The federal electric method analyzes complex maintenance tasks and applies time analysis in order to gauge the maintainability of equipment. The following are the four major steps:

1. Identification of principal parts
2. Determination of the failure rate of each part
3. Determination of the time required for the maintenance of each part

4. Computation of the expected maintenance time for the equipment by utilizing the information obtained in the first three steps

While the first two steps are concerned with routine maintenance, the ascertaining of time required to maintain each part forms the heart of this methodology. In order to ascertain maintenance time for each part of the equipment, the following seven actions are recognized. These actions are essential in order to restore broken equipment back to working condition.

1. *Localization.* The first step is concerned with pinpointing the location of the malfunction without using auxiliary test equipment.
2. *Isolation.* This step is concerned with determining the location of the malfunction by the use of the appropriate auxiliary test equipment, built-in test points, and so on.
3. *Disassembly.* Disassembly (full or partial) is essential in order to remove or replace a defective part(s) from a machine. Clearly, factors such as accessibility, ease of component removal, and so on play their usual roles here too.
4. *Interchange.* This process involves the substitution of a sound part in working condition for one or more that have failed.
5. *Reassembly.* As the name implies, this process involves restoring equipment to its original condition after disassembly.
6. *Alignment.* The various steps involved in this process are making adjustments, calibrations, and other checks and changes that have been made necessary due to repair action.
7. *Checkout.* This step involves the verification of the desired level of performance. It makes certain that the equipment has indeed been restored to its initial condition or any other condition that was destined for it.

The RCM Methodology

Reliability centered maintenance (RCM) was developed in the aviation industry in order to determine scheduled maintenance policies for civil aircraft. RCM emphasizes the role of reliability in focusing preventive maintenance activities on certain aspects. These aspects enable retention of the equipment's inherent design reliability. Clearly, this maintenance technique centers on reliability technology. The RCM philosophy was a result of efforts by industry, especially the airline industry (United Airlines, in particular) in the 1960s to undertake a complete reevaluation of preventive maintenance strategy. Since then, the importance of RCM has grown by leaps and bounds. For example, RCM specifications have been developed (MIL-STD, 1843, 1985), a course in RCM is offered by the Air Force Institute of Technology, and the Navy has published a handbook on RCM (S 9081-AB-GIB-010/MAINT, 1983).

The RCM methodology is a special case of Pareto analysis where resources are focused to solve the few yet vital problems that could cause serious system malfunction. It can be completely described by means of the following four unique features:

1. Preserve functions
2. Identify failure modes that can defeat the functions
3. Prioritize function needs (via failure modes)
4. Select applicable and effective preventive maintenance tasks for high-priority failure modes.

The Society of Automotive Engineers (SAE) Maintainability Standard

The SAE formulated a design for maintenance standard to be utilized early during the design stage of a new product, system, or machine (SAE J817-1, 1976). The SAE information report established a hierarchy of product effectiveness, defined serviceability, maintainability, reparability, and diagnostics. It related the attributes mentioned earlier to product effectiveness. The standard under discussion serves to establish a numerical value to rate an existing machine or a new conceptual machine. Maintenance is the primary criterion for evaluation. Based on certain requirements, the SAE index assigns point values to lubrication and maintenance items. Lubrication and maintenance are possible subjects to preset conditions such as location, accessibility, operation, and other miscellaneous factors. These requirements are clearly defined (as will be elaborated in the following pages). Each maintenance operation is described in detail and evaluated using conventional task analysis procedures. Each suboperation requiring the use of features such as location, accessibility, and so on is noted. Scores are assigned based on a preconceived system of scoring. The higher the score, the lower the maintainability of the machine and vice versa. Each requirement enumerated in the original standard is reproduced in the following tables briefly. It is to be noted that the degree of ease with which a requirement can be accomplished translates to a higher or lower score assignment.

A Design for X (DfX) Based Design for Maintenance Methodology

In this methodology, the most common and widely used maintenance operations are recorded and described in complete detail. Every maintenance operation is then subdivided into basic elemental tasks. Only a small fraction of all the tasks in the entire maintenance operation are actually responsible for performing effective maintenance. The remaining tasks include such actions as reaching for tools, grasping tools, and cleaning components prior to maintenance.

Incorporating Human Factors Into the Methodology

Maintenance workers, in most cases, need to adopt a particular posture and expend a requisite amount of energy in order to accomplish a particular maintenance task. There are several postures such as bending, squatting, and so on that are uncomfortable. An effective design methodology needs to take into account the different postures adopted by workers when performing maintenance tasks.

Format of the Design for Maintenance Index

The index under consideration will consist of three distinct sections. The first section is comprised of premaintenance actions such as slackening, tightening, and so on. The second section of the index will focus on evaluating the design variables directly affecting the actual maintenance processes such as lubrication, fitting, replacement, and so on. The third section takes into account all allowances that play a role in all maintenance operations. Human factor allowances have also been incorporated into the methodology as described earlier. These include allowances for postural requirements, manpower requirements, visual fatigue, and so on.

Developing a Numeric Index to Evaluate Ease of Maintenance

Most maintenance operations are comprised of three major tasks, namely disassembly, maintenance, and (re)assembly. Products that can do away with the need to disassemble a product prior to maintenance are often the easiest to maintain. An abridged version of a time-based numeric index for accessibility of joints and grooves has been presented in Table 4.1. A score

Table 4.1. Numeric Scores for Accessibility of Joints and Grooves for the Maintainability Index

<table>
<tr><td rowspan="6">Accessibility of joints and grooves</td><td rowspan="3">Dimensions</td><td rowspan="3">Length, breadth, depth, radius, angle made with surface</td><td>1.0</td><td>Shallow and broad fastener recesses, large and readily visible slot or recess in case of snap fits</td></tr>
<tr><td>1.6</td><td>Deep and narrow fastener recesses, obscure slot or recess in case of snap fits</td></tr>
<tr><td>2.0</td><td>Very deep and very narrow fastener recesses, slot for prying open snap fits difficult to locate</td></tr>
<tr><td rowspan="3">Location</td><td>On plane surface</td><td>1.0</td><td>Groove location allows easy access</td></tr>
<tr><td>On angular surface</td><td>1.6</td><td>Groove location is difficult to access; some manipulation required</td></tr>
<tr><td>In a slot</td><td>2.0</td><td>Groove location very difficult to access</td></tr>
</table>

of 73 time measurement units (TMUs) was assigned to this task, which corresponded to the time duration of approximately 2 seconds.

Maintenance operations are largely performed by people and often require the adoption of unnatural postures. This component of the index addresses the human factor aspect of the operation. A table of percentage multipliers for motion allowances is presented in Table 4.2.

The methodology for enhancing the disassemblability of a product is presented in Figure 4.1 by means of a decision-making flowchart. See Figure 2.2 in chapter 2.

A similar numeric index for assembly is presented in Table 4.3. Table 4.4 presents a similar index for an actual maintenance operation such as lubrication.

CASE STUDY ILLUSTRATING PRACTICALITY OF THE APPROACH

This section presents a case study to illustrate the lubrication process of a handheld drill rotor. A list of individual components for the drill is presented in Table 4.5.

The process illustrating the maintenance operation is presented in Table 4.6.

Table 4.2. Percentage Multipliers for Motions Allowance During Maintenance

Motions allowances	*Percentage multipliers (%)*
Normal motions	0
Limited motions	5
Awkward motions	5
Motions with confined limbs	10
Motions with confined body	10

Table 4.3. Numeric Index for Assembly Force During (Re)assembly Postmaintenance

Design attribute	*Design feature*	*Design parameters*	*Score*	*Interpretation*
Assembly force	Straight line motion without exertion of pressure	Push operations by hand	0.5	Little effort required
			1.0	Moderate effort required
			2.0	Large amount of effort required
	Straight line and twisting motion without pressure	Twisting and push operations by hand	1.0	Little effort required
			2.0	Moderate effort required
			4.0	Large amount of effort required

Table 4.4. Numeric Index for Lubrication

Lubrication	*Design feature*	*Score*	*Interpretation*
• Lubrication with brush • With oil can • With grease gun • Hand packing • Through access point • Through adapter	Lubrication between mating surfaces to minimize friction and prevent material loss	2/location 2/location 2/location 15/location 1/location 3/location	• Minor lubrication on surface or crevice • Minor lubrication in inaccessible areas • Minor greasing in inaccessible areas, slots, and so on • Major lubrication compensate material loss, and so on • Simple lubrication through access point • Simple lubrication requiring special equipment

Table 4.5. List of Components of a Handheld Drill

No	*Component name*	*Component material*	*Quantity*
1	Front or back screw	Copper	6
2	Middle screw	Copper	2
3	Bushing	Brass	1
4	Insulating washer	Plastic	1
5	Upper housing	Plastic	1
6	Lower housing	Plastic	1
7	Rotor	Mixed	1
8	Wire lead	Copper or plastic insulation	1

CONCLUSION

This chapter introduced readers to the importance of conserving natural resources for industrial use. Several examples of base metal and precious metal prices were cited as a reflection of supply and demand characteristics.

One of the principal means by which conservation can be achieved is through regular equipment maintenance. That way, equipment usage can be maximized, within the limits of technical applicability. To that end, this chapter familiarized readers with different concepts related to maintenance. Various approaches adopted by researchers, both reactive as well as proactive, were presented in as much detail as possible. Toward the end, this chapter illustrated how human factors can be effectively incorporated in order to aid in design modifications for a process that is largely labor intensive. The most significant advantage of

Table 4.6. Maintenance (Lubrication) Operation of a Drill Rotor

Task number	Task description *Lubrication of drill rotor*	Task total	Intersurface friction	Intersurface wedging	Material stiffness	Component size	Component weight	Component symmetry	Force exertion	Torque exertion	Dimensions	Location	Accuracy of tool placement	Posture allowance	Motions allowance	Manpower allowance	Visual Fatigue Allowance (%)
			(Dis)assembly force			Material handling			Tooling		Accessibility and positioning			Allowances			
1	Remove upper housing																
1a	Unscrew 1st of 6 front/back screws	15.65	2.5	–	–	2.0	2	0.8	–	2	1.6	2.0	2.0	–	–	–	5
1b	Unscrew 2nd of 6 front/back screws	15.65	2.5	–	–	2.0	2	0.8	–	2	1.6	2.0	2.0	–	–	–	5
1c	Unscrew 3rd of 6 front/back screws	15.65	2.5	–	–	2.0	2	0.8	–	2	1.6	2.0	2.0	–	–	–	5
1d	Unscrew 4th of 6 front/back screws	15.65	2.5	–	–	2.0	2	0.8	–	2	1.6	2.0	2.0	–	–	–	5
1e	Unscrew 5th of 6 front/back screws	15.65	2.5	–	–	2.0	2	0.8	–	2	1.6	2.0	2.0	–	–	–	5
1f	Unscrew 6th of 6 front/back screws	15.65	2.5	–	–	2.0	2	0.8	–	2	1.6	2.0	2.0	–	–	–	5
1g	Unscrew 1st of 2 middle screws	15.65	2.5	–	–	2.0	2	0.8	–	2	1.6	2.0	2.0	–	–	–	5
1h	Unscrew 2nd of 2 middle screws	15.65	2.5	–	–	2.0	2	0.8	–	2	1.6	2.0	2.0	–	–	–	5

4d	Screw 1st middle screw	15.65	2.5	–	–	2.0	2	0.8	–	2	1.6	2.0	2.0	–	–	–	5
4e	Screw 2nd middle screw	15.65	2.5	–	–	2.0	2	0.8	–	2	1.6	2.0	2.0	–	–	–	5
4f	Screw 1st front/back screw	15.65	2.5	–	–	2.0	2	0.8	–	2	1.6	2.0	2.0	–	–	–	5
4g	Screw 2nd front/back screw	15.65	2.5	–	–	2.0	2	0.8	–	2	1.6	2.0	2.0	–	–	–	5
4h	Screw 3rd front/back screw	15.65	2.5	–	–	2.0	2	0.8	–	2	1.6	2.0	2.0	–	–	–	5
4i	Screw 4th front/back screw	15.65	2.5	–	–	2.0	2	0.8	–	2	1.6	2.0	2.0	–	–	–	5
4j	Screw 5th front/back screw	15.65	2.5	–	–	2.0	2	0.8	–	2	1.6	2.0	2.0	–	–	–	5
4k	Screw 6th front/back screw	15.65	2.5	–	–	2.0	2	0.8	–	2	1.6	2.0	2.0	–	–	–	5
		294	Total time for maintenance operation: 2940 TMUs = 1.764 minutes.														

Total maintenance time: 1.764 minutes
Task #1 for maintenance analysis: Unscrewing/screwing back of various screws of upper housing.
Most feasible cost effective design solution: Use toggle type snap fits for upper housing in place of screws to reduce maintenance time.

Total disassembly time for complete 100% disassembly: 2.044 minutes (maintenance requires almost complete disassembly of drill).
Task #1 for disassembly analysis: Disassembly of rotor-bushing subassembly.
Most important design anomaly for disassembly: Force required wedging out subassembly of rotor and bushings.

Task #1 for assembly analysis: Inserting trigger assembly. Total assembly time: 1.30 minutes.

Conclusion: Most amount of time is spent in accessing the maintenance area. Too many fasteners hamper disassembly and assembly.

Table 4.6. (continued)

Task number	Task description *Lubrication of drill rotor*	Task total	Intersurface friction	Intersurface wedging	Material stiffness	Component size	Component weight	Component symmetry	Force exertion	Torque exertion	Dimensions	Location	Accuracy of tool placement	Posture allowance	Motions allowance	Manpower allowance	Visual Fatigue Allowance (%)
			(Dis)assembly force			Material handling			Tooling		Accessibility and positioning			Allowances			
1i	Pull out upper housing	8.70	–	1	–	3.5	2	1.4	1	–	1.0	1.6	1.6	–	–	–	1
2	Access drill rotor																
2a	Pull out bushing	10.50	1.0	–	–	2.0	2	0.8	1	–	1.0	1.0	1.2	–	–	–	5
2b	Pull out insulating washer	10.50	1.0	–	–	2.0	2	0.8	1	–	1.0	1.0	1.2	–	–	–	5
3	Lubricate rotor and clean housing																
3a	Lubrication: 2 locations	4.20	–	–	–	–	–	–	–	–	–	–	–	–	–	–	5
3b	Clean upper housing: 2	2.02	–	–	–	–	–	–	–	–	–	–	–	–	–	–	1
4	(Re)assemble drill																
4a	Reinstall washer	10.50	1.0	–	–	2.0	2	0.8	1	–	1.0	1.0	1.2	–	–	–	5
4b	Reinstall bushing	10.50	1.0	–	–	2.0	2	0.8	1	–	1.0	1.0	1.2	–	–	–	5
4c	Refit upper housing	8.70	–	1	–	3.5	2	1.4	1	–	1.0	1.6	1.6	–	–	–	1

this methodology is the fact that it is time-based. As a result, two distinct product designs can be readily compared based on time for maintenance and, therefore, the cost of maintenance.

REFERENCES

Gits, C. (1992). Design of maintenance concepts. *International Journal of Production Economics*, *24*(3), 217–226.

Harring, M. G., & Greenman, L. R. (1965). *Maintainability engineering*. Durham, NC: Duke University.

Herbaty, F. (1990). *Handbook of maintenance management cost effective practices* (2nd ed.). Park Ridge, NJ: Noyes Publication.

Imrhan, S. N. (1991). Workspace design for maintenance. In A. Mital & W. Karwowski (Eds.), *Workspace, equipment and tool design* (pp. 149–174). Amsterdam, Netherlands: Elsevier.

McDaniel, J. W., & Askrein, W. B. (1985). Report submitted to U.S. Air Force Aerospace Medical Research laboratory. Ohio: Wright Patterson AFB.

Society of Automotive Engineers. (1976). *SAE J817-1: Engineering design serviceability guidelines—Construction and industrial machinery—Serviceability definitions—Off-road machines*. Warrendale, PA: Society of Automotive Engineers.

Swanson, L. (2001). Linking maintenance strategies to performance. *International Journal of Production Economics*, *70*, 237–244.

CHAPTER 5

AN OPTIMIZATION APPROACH TO MAXIMIZE SERVICE QUALITY SUBJECT TO A GIVEN RESOURCE LEVEL FOR HIGHWAY INFRASTRUCTURE MAINTENANCE

Manoj K. Jha

Department of Civil Engineering
Morgan State University

Bimal Devkota

Department of Civil Engineering
Morgan State University

Bheem Kattel

Department of Industrial Engineering
Morgan State University

OVERVIEW

Highway infrastructure elements, such as roads, bridges, and tunnels, deteriorate rapidly due to continuous and heavy traffic flow, weather, and accidents. Their timely upkeep and maintenance is critical for maintaining an acceptable service level and performance quality. However, due to budgetary and other constraints, highway infrastructure elements are generally not well maintained, resulting in a less than acceptable level of service to the highway users. In this chapter we describe existing approaches to highway infrastructure maintenance conducted by the Baltimore City Department of Transportation. Further, some operations research techniques in optimizing the resource effectiveness and quality of highway infrastructure are presented.

INTRODUCTION

Since highways are critical to the U.S. economy, it is very important to maintain the minimum service level of highway infrastructure to ensure better mobility and safety to the users and meet demands during emergencies. Highway infrastructures should be able to facilitate the future adaptation to accommodate changing demand. Elements of highway infrastructure deteriorate over time and the cost of maintaining their useful life before they reach the threshold level of deterioration is enormous. Once the condition falls below the threshold level, maintenance is not sufficient to keep it in a serviceable condition. Replacement is the only option at that stage. There is a need for an effective and continuing inspection and maintenance program to keep the infrastructures functioning. With tight budgets, government agencies throughout the United States are even finding difficulties in performing periodic maintenance. The Department of Highway Safety and Motor Vehicles' revenue has been shrinking, while rising demand has kept the minimum level of services increasing over the years. In this chapter we discuss the highway infrastructure maintenance practices currently adopted by the city of Baltimore. We then present some operation research techniques to optimize the resource effectiveness and quality of highway infrastructure.

CITY OF BALTIMORE'S TRANSPORTATION INFRASTRUCTURE NETWORK

The transportation infrastructure network of the city of Baltimore primarily consists of pavements and bridges. The city manages 2,000 directional lane miles (a total of 5,000 lane miles) of roadways that includes 7 miles of the Interstate (see Figure 5.1). The Engineering and Construction Division is responsible for tracking pavement conditions and projects and preparing projects for roadway resurfacing. The Maintenance Division also prepares plans of proposed repairs based on known problem areas and citizens' complaints. The results of the pavement management would provide a treatment plan that would support contract and routine maintenance activities.

Lots of resources have been spent by the city on the pavement condition survey in 1999, 2004, and 2008. The 1999 survey data is available only in a Microsoft Access database. The 2004 survey data is available in the Hansen pavement management software (developed specifically for pavement management) for each 100-block segment that has a link to a Geographic Information System (GIS) map layer but not a direct link to a GIS viewer. For the 2008 data, 100-block segments created in 2004 were related to the current city centerline file. The results have been displayed in the GIS and also used by Axiom Decision System (a program also developed for pavement management) to perform modeling for optimization. The Pavement Condition Index (PCI) of 0 to 100 is calculated based on the pavers

Figure 5.1. City of Baltimore street network.
Source: Courtesy of the City of Baltimore Department of Transportation.

program developed by the Army Corp of Engineers, with 100 being the best. A PCI of more than 70 is considered as an acceptable condition and less than 70 is considered as a substandard condition. Currently 59% of the city's pavement is in acceptable condition and 41% of the pavement is in substandard condition. Besides a pavement condition survey, the city has a historical record of the pavement, which is very critical in determining appropriate treatment based on the subbase construction of the roadway. Axiom's process used to create a maintenance optimization model relies on a genetic algorithm following both deterministic and probabilistic approaches.

Bridges

Baltimore has the responsibility of the maintenance and rehabilitation of 299 bridges within the city limits and watershed bridges in Baltimore County (see Figure 5.2). All bridges are inspected biannually since 1987. A small percentage of the bridges whose condition warrants rerating based on the conditions are inspected annually. The bridges are included in the Structural Inventory and Appraisal System (SAIS); conditions are noted on a plan and recommendations are prepared for the repair and maintenance of each bridge. All these data are loaded in a bridge management software program called PONTIS, owned by the Maryland State Highway Administration. After inspection, bridge sufficiency rate (BSR) is calculated from 0 to 100, with 100 being the best. Under the federal highway policy, bridges with a BSR greater than 80 do not require any maintenance, bridges with a BSR between 50 and 80 need rehabilitation, and bridges with a BSR less than 50 need to be replaced. By querying the PONTIS data, a user can access the elements of the bridge to be rehabilitated or replaced, the type of repairs recommended, and the cost of repair for the particular element or other elements of the bridge. Most bridge repair work is performed under contract. Minor repair and maintenance work is assigned to the Maintenance Division. A preventive maintenance program is currently being developed using the SAIS program.

RESOURCE EFFECTIVENESS AND SERVICE QUALITY

Resource effectiveness (RE) and service quality (Q) can be graphically illustrated as shown in Figures 5.3 and 5.4, respectively. Figure 5.3 shows that a higher level of resource effectiveness is generally needed to improve upon quality. After a certain threshold is reached, no further improvement in quality can be achieved regardless of the increased resource allocation. In other words, the optimal value of quality, say Q*, is insensitive to resource allocation. The resource effectiveness corresponding to Q* is said to be optimal, designated as RE* in Figure 5.3.

Figure 5.4 illustrates how different levels of service quality can be achieved for the same level of resource effectiveness, which implies that service quality can be optimized for a given level of resource. Thus, an optimization formulation can be developed to maximize Q subject to a given RE.

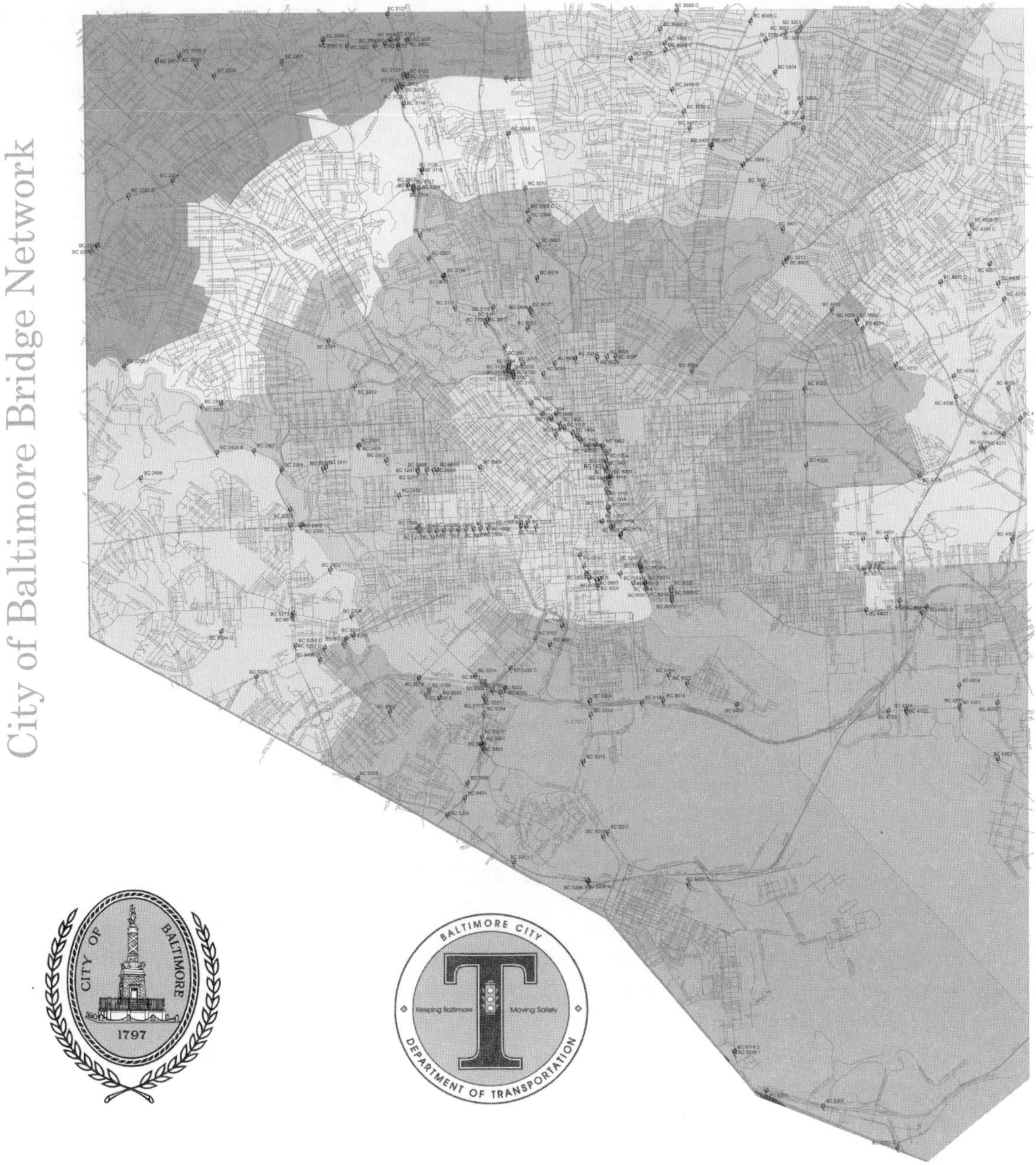

Figure 5.2. City of Baltimore road and bridge network.
Source: Courtesy of the City of Baltimore Department of Transportation.

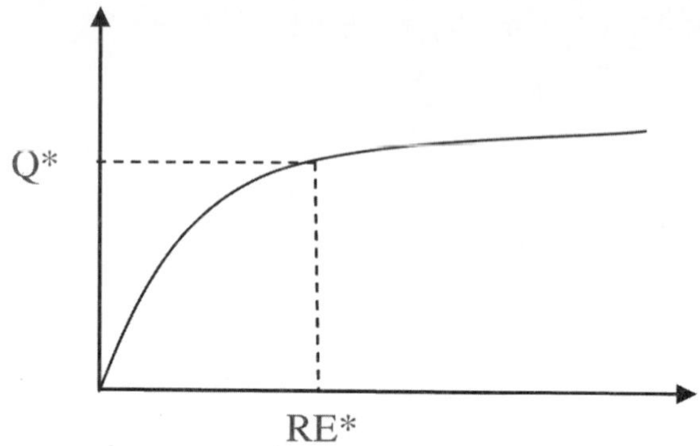

Figure 5.3. Graphical illustration of resource effectiveness (RE) versus service quality (Q).

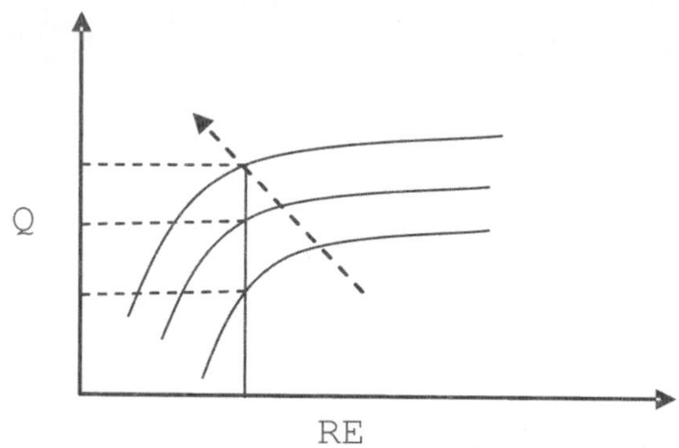

Figure 5.4. Different levels of quality for a given resource effectiveness (RE).

AN OPTIMIZATION APPROACH

The application of operations research (OR)-based optimization techniques in various infrastructure maintenance optimization problems is not new (Jha & Abdullah, 2006a). One needs to investigate the nature of the problem on hand, identify appropriate decision variables, and develop a mathematical optimization problem. A solution procedure needs to be devised based on the nature of the mathematical optimization problem. In most of the real-world problems it is difficult to employ exact procedures to obtain true optimal results. Therefore, numerical or approximate solution procedures using artificial intelligence (AI)-based algorithms are often employed (see Jha & Schonfeld, 2004; Kim, Jha, & Son, 2005).

Review of Some OR Techniques Employed in Infrastructure Maintenance Optimization

Infrastructure maintenance optimization problems have received widespread attention in the last 30 years primarily due to the recognition of the fact that the highway infrastructure was deteriorating rapidly and new and innovative methods were needed for maintenance. Golabi, Kulkarni, and Way (1982) developed a pavement maintenance optimization model using Markovian property—a sequential decision process whose description is available in Jha and Abdullah (2006a). This optimization formulation was solved using a linear programming approach. Jha and Abdullah (2006a) extended Golabi et al.'s work to roadside assets (such as guardrails, signs, and luminaries) and developed a maintenance optimization model for those assets. They proposed a genetic algorithm to solve the developed optimization formulation. Jha and his fellow researchers, in their subsequent works (Jha & Abdullah, 2006a, 2006b, 2008; Jha, Kepaptsoglou, Karlaftis, & Abdullah, 2006, 2008), successively refined the maintenance optimization formulation by introducing probabilistic, bilevel, and multiobjective approaches. In one of their recent works (Jha, Chacha, Udenta, & Abdullah, 2009), they introduced a capacitated arc routing problem for maintenance inspection by considering workshift and overtime limit constraints.

Maji and Jha (2007) developed a highway infrastructure maintenance schedule for roadside assets (such as guardrails and ground-mounted signs) subject to a budget constraint. The following is a developed optimization formulation:

$$\text{Min} \qquad TMC = \sum_{m=1}^{N} \sum_{n=1}^{T} x_{mn} \times pmc_{mn} \qquad (1a)$$

subject to

$$pc_{mn} \geq tc_{m} \qquad \forall\, m = 1, 2, \ldots, N \text{ and } \forall\, n = 1, 2, \ldots, T \qquad (1b)$$

$$\sum_{m=1}^{N} x_{mn} \times pmc_{mn} \leq b_{n} \qquad \forall\, n = 1, 2, \ldots, T \qquad (1c)$$

$$x_{mn} = 0, 1, \qquad (1d)$$

where

TMC = total maintenance cost

N = number of highway infrastructure elements

T = planning horizon or time period (usually in years)

pmc = present maintenance cost

pc = present condition of the infrastructure element

tc = threshold condition of the infrastructure element

b_n = budget allocation for the nth fiscal year of the design period

$$x_{mn} = \begin{cases} 1 & \text{if element } m \text{ maintained in the } n^{\text{th}} \text{ year} \\ 0 & \text{otherwise} \end{cases}$$

The first condition of the objective function (equation 1b) checks the condition of all elements each year with their respective threshold condition and the second condition (equation 1c) checks the maintenance cost with the budget allocation for each fiscal year. The optimization problem was solved using a genetic algorithm procedure, details of which can be found in Maji and Jha (2007).

Optimization Formulation

Here we develop an optimization formulation for maximizing resource effectiveness (which, in turn, maximizes service quality). It is assumed that a logarithmic relationship exists between resource effectiveness and service quality (see equation 2) and the service quality Q can be represented as a combined measure of labor (L), equipment (E), and material (M).

$$Q = \ln(RE) \qquad (2)$$

In order to express Q as a combined measure of labor (L), equipment (E), and material (M), we can use the well-known Cobb and Douglas (1928) production function as follows:

$$Q = AL^{\alpha}E^{\beta}M^{\gamma} \qquad (3),$$

where A is a constant and α, β, and γ are coefficients such that their values reside between 0 and 1. Also, $\alpha + \beta + \gamma = 1$, which implies a constant return to scale and that the Cobb and Douglas (1928) function is homogeneous. This means that doubling of all three inputs, L, E, and M (at the same time), results in a doubling of the output Q. The coefficients α, β, and γ represent the respective partial "output elasticities" of the service quality function Q, meaning that they measure the percentage increase in service quality resulting from a percentage increase in the particular input factor.

Thus, the optimization formulation for maximizing resource effectiveness R can be expressed as

$$\text{Max} \quad R(L, E, M) \qquad (4),$$

which is subject to the following:

$$R = AL^{\alpha}E^{\beta}M^{\gamma} \qquad (5a)$$

$$B(L, E, M) = lL + eE + mM \leq B \qquad (5b)$$

$$\alpha + \beta + \gamma = 1 \qquad (5c)$$

$$A, \alpha, \beta, \gamma \geq 0 \qquad (5d)$$

Please note that for notational convenience the resource effectiveness *RE* has been expressed as *R* in the previous formulation, which is a function of *L*, *E*, and *M*. In equation 5b, *l*, *e*, and *m* are unit costs of labor, equipment, and material, respectively; *B* is the budget function; and *M* is the available budget.

The previous optimization problem can be solved by relatively easier techniques (such as using the Lagrangian Multiplier), and an exact solution to this problem may be obtained by calculus-based methods. The implication to the formulation in the infrastructure maintenance context is that one has to obtain the optimum combination of labor, equipment, and material in order to maximize resource effectiveness. It will be desirable to obtain historical maintenance cost data for the city of Baltimore, with which we can run different scenarios to optimize *RE* and also investigate the sensitivities of *L*, *E*, and *M* on *RE*. In this chapter, however, we will perform a simple numerical example.

SOLUTION PROCEDURE AND A NUMERICAL EXAMPLE

A well-known technique to solve the previous optimization problem is one that uses a Lagrangian Multiplier. Using the Lagrangian Multiplier the optimization problem can be expressed as

$$R' = R - \lambda[(lL + eE + mM) - B] \qquad (6),$$

where λ is the Lagrangian Multiplier, also known as marginal value of shadow price. Specifically, the value of the Lagrangian Multiplier is the rate at which the optimal value of the function *R* changes if constraints are changed.

Equation 5 is solved by taking the first order conditions and setting them equal to zero.

A Numerical Example

We present a simple numerical example of resource maximization using only two variables of labor and equipment for a highway infrastructure maintenance problem. In this example, we assume $R = 12L^{2/3}E^{1/2}$. We further assume that cost of labor is \$40 per hour and cost of equipment is \$30 per hour and the available budget is \$70,000 to carry out the maintenance operation. Solving this problem using the approach presented here, we obtain the total labor cost to be \$10,000 and total equipment cost to be \$10,000, resulting in a *R* value of \$574,356 and in turn an *Q* value of 13.26.

While this example is a simple two-dimensional application of the earlier approach for resource effectiveness optimization, more complex scenarios with additional variables and real-world data can be solved in future works.

CONCLUSIONS AND FUTURE WORKS

In this chapter, we developed an optimization problem for maximizing resource effectiveness for highway infrastructure maintenance. Various sensitivity analyses using real-world data will be conducted in future works.

ACKNOWLEDGMENTS

This work was carried out at the Center for Advanced Transportation and Infrastructure Engineering Research (http://www.eng.morgan.edu/~catier), Morgan State University. The authors wish to thank city of Baltimore Department of Transportation for supplying the road and bridge network data and maps.

REFERENCES

Cobb, C. W., & Douglas, P. H. (1928). A theory of production. *American Economic Review, 18*(1; Suppl., *Papers and proceedings of the 40th annual meeting of the American Economic Association*), 139–165.

Golabi, K., Kulkarni, R. B., & Way, G. B. (1982). A statewide pavement management system. *Interfaces, 12*(6), 5–21.

Jha, M. K., & Abdullah, J. (2006a). A Markovian approach for optimizing highway life-cycle with genetic algorithms by considering maintenance of roadside appurtenances. *Journal of the Franklin Institute, 343*, 404–419.

Jha, M. K., & Abdullah, J. (2006b). A probabilistic approach to maintenance repair and rehabilitation (MR&R) of roadside features. *Proceedings of the 14th Pan-American Conference on Traffic and Transportation Engineering*, Las Palmas de Gran Canaria, Spain.

Jha, M. K., & Abdullah, J. (2008). A probabilistic bi-level optimization approach to highway infrastructure maintenance in urban areas. In E. Taniguchi & R. Thompson (Eds.), *Innovations in city logistics* (pp. 571–584). New York, NY: Nova Science Publishers.

Jha, M. K., Chacha, S., Udenta, F., & Abdullah, J. (2009). Formulation and solution algorithms for highway infrastructure maintenance optimization with work-shift and overtime limit constraints. *Proceedings of the 6th International Conference on City Logistics*, Puerto Vallarta, Mexico.

Jha, M. K., Kepaptsoglou, K., Karlaftis, M., & Abdullah, J. (2006). A genetic algorithms-based decision support system for transportation infrastructure management in urban areas. In E. Taniguchi & R. Thompson (Eds.), *Recent advances in city logistics* (pp. 509–523). Oxford, England: Elsevier.

Jha, M. K., Kepaptsoglou, K., Karlaftis, M., & Abdullah, J. (2008). A bilevel optimization model for large-scale highway infrastructure maintenance inspection and scheduling. *10th International Conference on Application of Advanced Technologies in Transportation*, Athens, Greece, May 2008.

Jha, M. K., & Schonfeld, P. (2004). A highway alignment optimization model using geographic information systems. *Transportation Research, Part A*, *38*(6), 455–481.

Kim, E., Jha, M. K., & Son, B. (2005). Improving the computational efficiency of highway alignment optimization models through a stepwise genetic algorithms approach. *Transportation Research, Part B*, *39*(4), 339–360.

Maji, A., & Jha, M. K. (2007). Modeling highway infrastructure maintenance schedule with budget constraint. *Journal of the Transportation Research Board*, 19–26.

PART IV

HUMANS

CHAPTER 6

HUMANS AND TECHNOLOGIES IN SYMBIOTIC RELATIONSHIPS: A JOINT OPTIMIZATION FRAMEWORK FOR ENHANCING EFFECTIVENESS

Anil Mital
Industrial Engineering
University of Cincinnati

Arunkumar Pennathur
Industrial Engineering
University of Texas

Rajeev Senapati
Industrial Engineering
University of Texas

INTRODUCTION

Advanced technologies and humans are two critical resources for any industry and play important roles in effective and efficient functioning of any modern production or service system. It was widely believed in the 1980s that complete automation of industries through design concepts such as lights-out factories (fully automated factories that do not require any human presence on-site) would eliminate human influence from industries and improve industrial productivity (Bulinger & Warnecke, 1985). Humans were viewed as error-prone and inefficient. However, completely automated factories have not succeeded, except in a few isolated cases. Indeed, human operators are playing, and will continue to play, important roles in enhancing industrial productivity (Mital, 1997). However, technology and automation are also important resources in enhancing industrial productivity.

Given that both humans and technologies are important resources in the modern industrial workplace, the key challenge for a system designer, then, is to jointly optimize human and technology resources. Humans and technologies both have limitations and capabilities that need to be taken into account when designing to optimize the human–automation interaction.

This chapter presents a framework for jointly optimizing human and technology resources for enhancing resource utilization and productivity. The limitations and capabilities of humans and technologies are briefly presented first. Then, a framework for jointly optimizing the human–technology interactions is presented with three distinct steps in the framework: user needs analysis, function allocation for system design, and guidelines for designing displays and controls found in automated devices.

HUMAN AND TECHNOLOGY CAPABILITIES AND LIMITATIONS: A BRIEF OVERVIEW

Humans are capable of making decisions under unpredictable situations, such as technical breakdowns. They can effectively control the system under such situations. Humans are more flexible and can easily adapt to deal with any changes in the system. But humans have some limitations. Humans are very unpredictable and are believed to be the cause of uncertainty; humans have great difficulty in providing the required quality, uniformity, cleanliness, reliability, and repeatability in any work (Mital & Pennathur, 2004). Humans also have a small working memory and are unable to retrieve past information.

Advanced technology is used in industries to overcome the limitations of the human operators. Technology can be used to perform repetitive routine tasks and aid in performing complex cognitive tasks. Advanced technology also affords large data storage and processing and enables multitasking. But, advanced technology has limitations as well. Technology causes a deskilling of the human operators. In systems with extensive use of technology, the role of the human is just to supervise the system; this causes the loss of skills in the human operators. Technology is also inflexible and cannot adapt to new systems (Mital & Pennathur, 2004). There are task-device cycles associated with advanced technology. By the time workers learn to use a technological artifact for performing their work, new technologies and designs impose additional learning burdens on the human. Technologies that workers use in the workplace oftentimes also require significant workarounds from the humans because of poor usability.

A lack of awareness of the different modes a certain machine has is a critical problem with advanced technology. Automated systems oftentimes provide very little or no indication about their current working status or what task follows. Because of this, human operators are unaware of the status of the machine's task, which causes them to have to perform significant cognitive work to infer the current and future status of the machine. Lack of mode awareness creates a gap between the operator's expectations of machine status (mental model) and the actual machine status, due to the fact that the operators often are unable to anticipate the behavior of the automated machine. This results in automation surprises that human operators have to deal with. A poor mental model match, low system observability, and highly dynamic or non-routine situations all contribute to automation surprises due to advanced technology (Sarter & Woods, 1997).

THE HUMAN–MACHINE SYSTEM FRAMEWORK FOR INTERACTION DESIGN

Traditionally, system designers have accounted for human limitations and capabilities by considering the human operator as an information processor having sensory and motor capabilities and limitations. The key elements to the efficient and error-free functioning of a human–machine system are the provision of information to human operators in the system and the provision for control of the system by humans.

Displays provide information about the machine or the system to human operators, and controls enable human operators to take actions and change machine or system states (conditions). Operator feedback is obtained through interaction with the controls (e.g., tactile sensing). Thus, in the classical view, human interaction with automation is mediated through displays and controls for a two-way exchange of information.

The recent view of the human–machine system, resulting from advances in computerized information systems, sees the human operator as a supervisory controller responsible for supervisory functions such as planning, teaching, monitoring, intervening, learning, and so on (Sheridan, 1987). Even though, in such a view, the human operator has a changed role, displays and controls still provide the fundamental medium for human interaction with the system.

Indeed, properly designed displays and controls are fundamental to the efficient and error-free functioning of manufacturing systems. The human factor of engineering, which we define as the study of issues involved in the application of technology to assist the human element in work and in the workplace, provides recommendations for interface design based on research in human sensory and motor capabilities and limitations.

Even though displays and controls and their effective design are fundamental to the efficient and error-free operation of the system, a number of important activities need to be carried out before one can think of displays and controls. These activities stem from the central need to build systems to suit human limitations and capabilities. Throughout this chapter, we focus on the essence of the human-centered interface design approach. We first present elements of this approach and contrast it briefly with a system-centered interface design approach. It is recommended that this concept of human-centered design guides the designer at both the system and the nuts-and-bolts design level (i.e., machine displays and controls).

The System-Centered Design Approach

The system-centered design approach, as the name suggests, analyzes the system currently in use, designs and specifies the new system, builds and tests the new system, and delivers the system and makes minor changes to it (Figure 6.1). The focus is on the goals of the system and the goals of the organization within which the system performs. Designers following this approach fail to consider system users. Hence, users have to remember too much information. These systems are intolerant of minor user errors and are confusing to new users. Often, such systems do not provide the functions users want, and users have to perform tasks in undesirable ways. New systems designed the system-centered way have also been shown to cause unacceptable changes to the structure and practices in entire organizations (Booth, 1989).

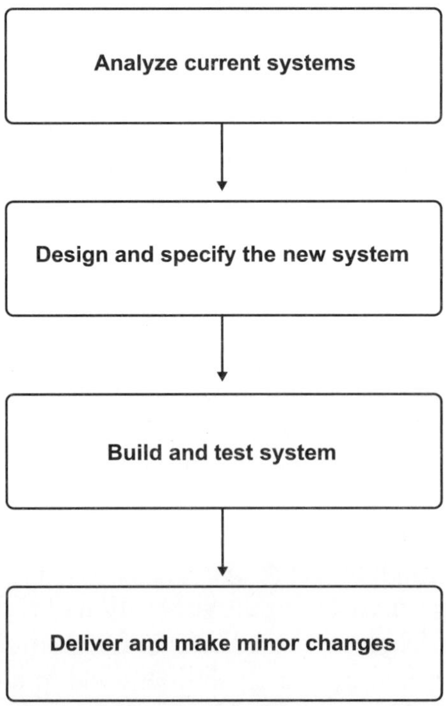

Figure 6.1. System-centered approach to human–machine interaction design.

The Human-Centered Design Approach

The human-centered design approach to human–machine interaction, unlike the system-centered approach, puts human capabilities and limitations ahead of the system goals. In other words, the entire system is built around the user of the system. This approach has been variously called the "usability engineering approach," the "user-centered approach," or the "anthropocentric approach to production systems." Figure 6.2 provides our conception of the human-centered approach to interface design.

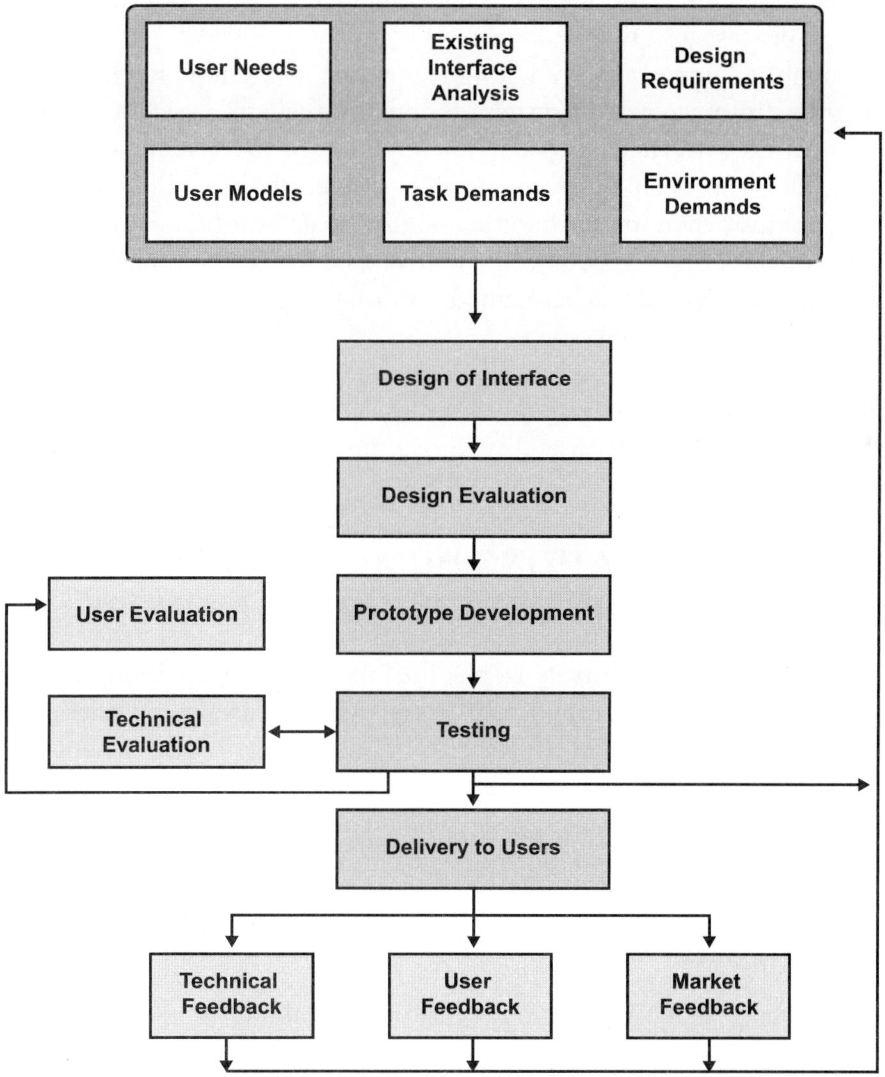

Figure 6.2. Human-centered approach to human–machine interaction design.

The first step in this design approach is information collection. Information about user needs, information about user cognitive and mental models, information on task demands, information on the environment in which the users have to perform, information on the existing interface between the human operator (the user of the system) and the machine(s), requirements of the design, and so on are some of the more important variables about which information is collected. This information is then used in the detailed design of the new interface.

The design is then evaluated. Prototype development and testing is conducted. User testing and evaluation of the prototype, the other important characteristic of this design process, is then carried out. This results in new input to the design of the interface, making the entire design process iterative in nature.

Even though the human-centered design approach is intended to take human capabilities and limitations into account in system design and make the system usable, there are a number of difficulties with this approach. The usability of the system is only as good as its usability goals. One approach to overcome this problem is to include users when setting usability goals, not just when measuring the usability goals. Another common difficulty with this approach is the lack of provision for qualitative data for design refinement. There is also the drawback that this approach is best suited for designing new systems and that it is not as effective for redesign of existing systems.

Despite these limitations, the human-centered design approach merits consideration from designers because it proactively takes the user of the product into the system design process and, as a result, engineers usability into a human–machine interface.

SYSTEMATIZING HUMAN–MACHINE INTERACTION DESIGN: THE FRAMEWORK

Even though the displays and controls are the final means of information exchange between humans and machines in a system, the actual design of the hardware and software for displays and controls comes last in order in the process of optimizing human–machine interaction. The other key steps in this process include user needs analysis, task analysis, situation analysis, and function allocation decisions, after which the modes of information presentation and control can be decided. In the following sections, each of these steps is briefly discussed.

User Needs Analysis

The goal of user needs analysis is to collect information about users and incorporate it into the design process for a better design of human–machine interaction. User needs analysis typically involves the characterization of the user, the characterization of the task the user performs, and the characterization of the situation under which the user has to perform the task. In the next few sections, we provide guidelines and methods for performing each of these three activities prior to designing the system.

Characterization of the User

Table 6.1 provides a user characterization checklist. Included in this checklist are questions to elicit information about the users, information about the users' jobs, information about the users' backgrounds, information about usage constraints, and information about the users' personal preferences and traits. The goal of collecting such information is to use the information in designing a usable system.

Table 6.1. User Characteristics Checklist

Data about users	What is the target user group?
	What proportions of users are male and what proportion are female?
	What is the average age and age range of users?
	What are the cultural characteristics of users?
Data about job	What is the role of the user (job description)?
	What are the main activities in the job?
	What are the main responsibilities of the user?
	What is the reporting structure for the user?
	What is the reward structure for the user?
	What are user schedules?
	What is the quality of output from the user?
	What is the turnover rate of the user?
Data about user background	What is the education/knowledge/experience of the user relevant to the job?
	What are the relevant skills possessed by the user?
	What relevant training have the users undergone?
Data about usage constraints	Do users voluntarily use the current equipment?
	What are the motivators and demotivators for use?
Data about user personal preferences and traits	What is the learning style of the user?
	What is the interaction style of the user?
	What is the aesthetic preference of the user?
	What are the personality traits of the user?
	What are the physical traits of the user?

Adapted from Booth (1989).

Characterization of the Task

Characterization of the tasks users have to perform to attain system goals is done through task analysis. Task analysis is defined as the formal study of what a human operator (or a team of operators) is required to do to achieve a system goal (Kirwan & Ainsworth, 1992). The actions or the cognitive processes involved in achieving the system goal are assessed.

The key questions to ask when performing task analysis activities are shown in Table 6.2. The task analysis methodology finds use for designers at all stages in the life cycle of a system, from initial conception through the preliminary and detailed design phases, then to the prototype and actual product developments, and finally to the storage and demolition stage. While many different task analysis techniques exist to suit the different design requirements in systems, our primary focus here is on techniques that help in designing the human–machine interaction.

The key issues involved in designing human interaction with automated equipment are assessing the types of information that human operators will need to understand the current system status and requirements and the types of output that human operators will have to make to control the system. Table 6.3 provides a summary of the important activities involved in the process of interface design and the corresponding task analysis technique to use in designing this activity. The reader should refer to Kirwan and Ainsworth (1992) or other articles on task analysis for a detailed discussion of the different task analysis techniques. The following are some important techniques used for task interaction design:

Task decomposition. This is a method used to state the tasks involved in terms of information content and the actions and feedback required of the operator. When a list of activities and the tasks involved has been generated, task decomposition can be used to systematically expand on the task descriptions. The various steps involved in task decomposition are presented in Figure 6.3.

Decision-action diagram. This is one of the most commonly used tools for decision making (Mital, Morotwala, Kulkarni, Sinclair, & Siemieniuch, 1994). The decision-action diagram sequentially proceeds through a series of questions (representing decisions) and possible yes or no answers (representing actions that can be taken). The questions are represented as diamonds, and the possible alternatives are labeled on the exit lines from the diamond. A thorough knowledge of the system components and the possible outcomes of making decisions about system components are essential for constructing complete and representative decision-action diagrams.

Tabletop analysis. As the name implies, this is a technique through which experts knowledgeable about a system discuss specific system characteristics. In the context of interaction design, this task analysis methodology, used for checking if the information flows as identified during the initial task analysis and task description, is adequate for successful task completion. Tabletop analysis, hence, typically follows the initial hierarchy or other forms of task analysis that yield task descriptions and provide information input for the decomposition of the tasks.

Table 6.2. Checklist for Task Analysis Activities

Goals	*What are the important goals and supporting tasks?*
For every important task:	
Intrinsic of the task	What is the task?
	What are the inputs and outputs of the task?
	What is the transformation process (inputs to outputs)?
	What are the operational procedures?
	What are the operational patterns?
	What are the decision points?
	What problems need solving?
	What planning is needed?
	What is the terminology used for task specification?
	What is the equipment used?
Task dependency and criticality	What are the dependency relationships between the current task and the other tasks and systems?
	What are the concurrently occurring effects?
	What is the criticality of the task?
Current user problems	What are the current user problems in performing this task?
Performance criteria	What is speed?
	What is accuracy?
	What is quality?
Task criteria	What is the sequence of actions?
	What is the frequency of actions?
	What is the importance of actions?
	What are the functional relationships between actions?
	What is the availability of functions?
	What is the flexibility of operations?
User discretion	Can the user control or determine pace?
	Can the user control or determine priority?
	Can the user control or determine procedure?
Task demands	What are the physical demands?
	What are the perceptual demands?
	What are the cognitive demands?
	What are the environmental demands?
	What are the health and safety requirements?

Adapted from Booth (1989).

Table 6.3. Summary of Task Analysis Activities and Methods Involved in Interaction Design

Activity	*Task analysis method*
Gathering task analysis representing the activities within the task	Hierarchical task analysis Activity sampling
Stating required information, actions, and feedback	Work study Task decomposition Decision–action diagrams
Checking adequacy of provisions for information flows for successful completion of task	Tabletop analysis Simulation Walkthrough and talkthrough Operator modifications surveys Coding consistency surveys
Identifying links between attributes (total system check) to ensure system success	Link analysis Petri nets Mock-ups Simulator trials
Provide detailed design recommendations	Person specification Ergonomics checklist

Modified from Kirwan and Ainsworth (1992).

Walkthrough and talkthrough analysis. These analyses involve operators and other individuals having operational experience with the system so they can walk and talk the analyst through observable task components of a system in real time. A walkthrough is normally achieved in a completely operational system or in a simulated setting. A talkthrough can be performed even without a simulation of the system; the only requirements are drawings and other system-specific documentation with enough information to enable the analysts to set system and task boundaries while performing the talkthrough analysis. For more information on walkthrough and talkthrough analyses, refer to Meister (1986).

Operator modification surveys. These surveys are performed to gather input from the actual users of the system (i.e., the operators) to check if there will be difficulties in using the system and identify these difficulties. This checking of the adequacy of the interface design of the system from the users' perspective is done through surveys conducted on similar operational systems. In general, operators and other users of systems maintain and provide

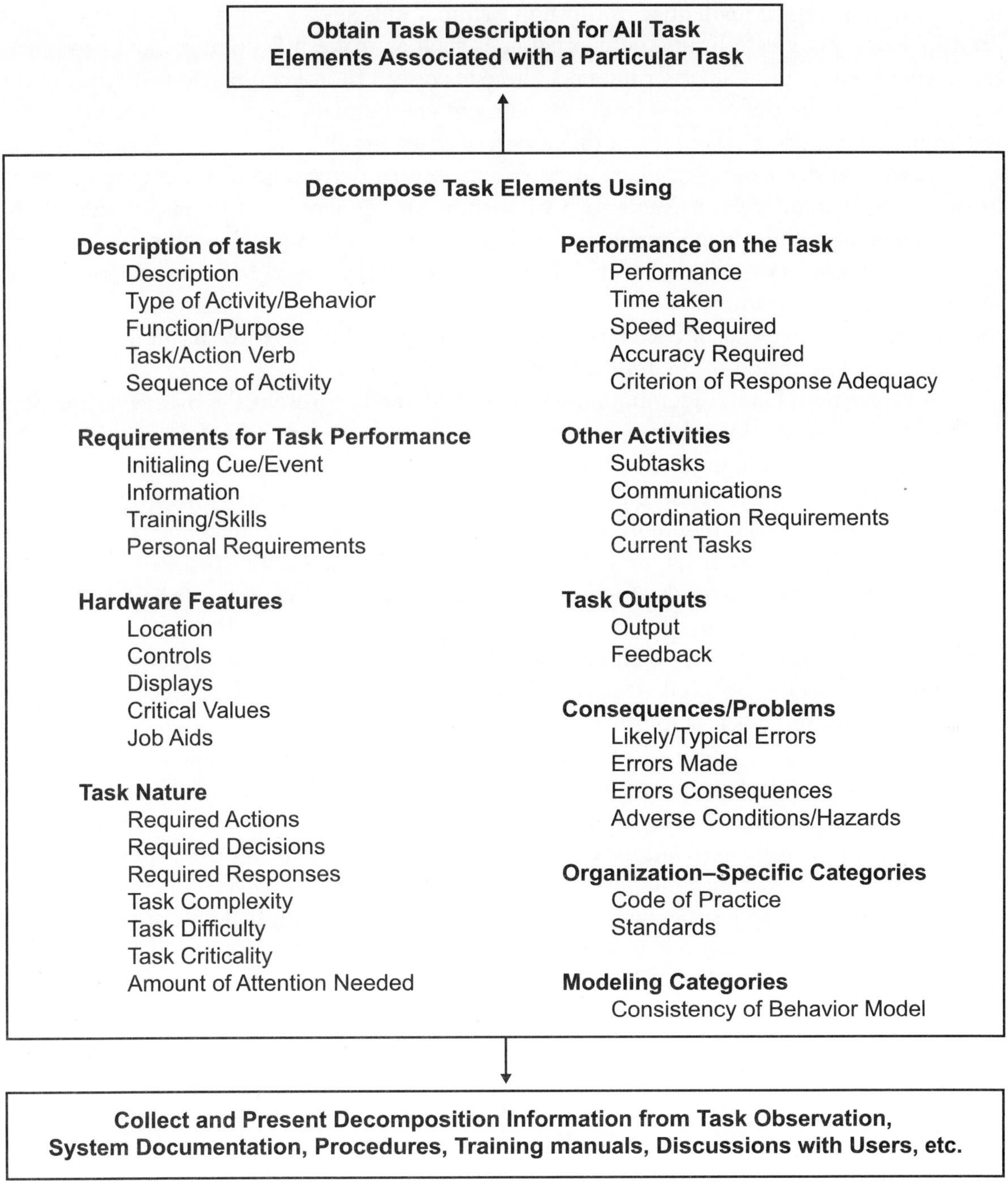

Figure 6.3. The task decomposition process.

information on design inadequacies through memory aids such as their own labels or signs on displays to mark safe limits, perceptual cues such as makeshift pointers, and organizational cues such as lining up instruments. These makeshift modifications done by the operators indicate design deficiencies in the system and can be planned for and included in the redesign of the existing system or in the design of a new system.

Coding consistency surveys. These surveys are used to determine if the coding schemes in use in the system are consistent with what the codes generally mean, and if and where additional coding is needed. The recommendation when performing the coding consistency surveys is to record a description of the location of the item, a description of the coding used for that item (e.g., an intermittent siren sound), a description of any other coding schemes for that item (e.g., an intermittent siren sound accompanied by a yellow flashing light), and the complete description of the function being coded.

Link analysis. This is a technique used to identify and represent the nature, frequency, or the importance of the relationship or links existing between individual operators and some portion of the system (Chapanis, 1976). Link analysis begins by first collecting information about the system components used during task performance. This information is then used to develop a complete list of links between individual system elements. The links that are established are then diagramed and ranked for importance. The order of importance may be determined based on the frequency of activity between two links or based on other appropriate measures decided by the system expert. The nature of the links to be studied (is it a movement of attention or position between parts of the systems?) and the level of detail to include in defining each link are important factors that determine the overall structure and usefulness of the links established.

Simulator analysis. The goal of simulation studies is to replicate and observe system (including operator and operating environment systems) performance while making the performance environment as representative and close to the real-time environment as possible. Different forms of simulations exist depending on the platform or the simulator used for the simulation. Examples include a simple paper-and-pencil simulation, a mock-up of a system that may or may not be dynamic, and a dynamic simulation that will respond in real time. Whatever the method of simulation used, the key consideration in simulation studies is the trade-off between the fidelity of the simulation (deciding the features of the system that need fidelity is an issue, too) and the cost involved in building high-fidelity simulations.

Person specification. The goal of person specification is to detail the key physical and mental capabilities, qualification and personality traits, and experience that is required of the operator to perform specified tasks. A widely used technique for person specification is the position analysis questionnaire. Position analysis questionnaires require the operator to identify their specified tasks and jobs, the information input, the mental processes, the work output, the context of the job, the relationship with other personnel in the system, and any other relevant job characteristics. Using the responses from the operators, the skill content of tasks and jobs can be determined and can help in designing personnel selection and training programs to ensure optimal human–machine interaction.

Ergonomics checklists. These checklists are generally used to ascertain if a particular system meets ergonomic principles and criteria. Ergonomics checklists can check for subjective or objective information and can cover issues ranging from overall system design to the design of individual equipment. Checklists can also range in detail from the broad ergonomic aspects to the minute detail. Table 6.4 provides an example of a checklist for equipment operation. Important ergonomics checklists include the ones by Woodson (1981; Woodson, Tillman, & Tillman, 1991); the MIL-STD 1472C (Department of Defense, 1981), which covers equipment design (written primarily for military equipment but can be used as a guide to develop checklists); the EPRI NP-2360 (Seminara, 1949), which is a checklist for maintenance activities in any large-scale systems; the NUREG0700 (Nuclear Regulatory Commission, 1981), which is a comprehensive checklist for control room designs; the health, safety, and environment (HSE) checklist (HSE, 1989), which deals with industrial safety and human error; and the numerous checklists for cathode ray tube (CRT) displays and visual display units (VDU) (Blackman, Gertman, & Gilmore, 1983; Cakir, Hart, & Stewart, 1980).

Characterization of the Situation

The external environment in which the system functions can influence the human–system interaction performance. Table 6.5 provides a checklist for the most commonly encountered situations for which the system analyst must obtain answers. The checklist attempts to provide for these situations in design.

Allocation of Functions

In designing the human–machine interaction, once comprehensive information about the users and the activities and tasks these users will perform is known, the activities and tasks need to be assigned either to the humans or to the machines. Allocation of functions is a necessary first step before any further design of the interaction in the human–machine system can be carried out.

The need for solving the function allocation problems directly stems from the need to decide the extent of automation of activities. This is so because, in the present-day manufacturing scenario, the decision is no longer whether or not to automate tasks but to what extent and how.

The function allocation problem is perhaps as old as the Industrial Revolution itself. Fitts's list, conceived in 1951 (Table 6.6), was the first major effort to resolve the function allocation problem. Further research on function allocation has permitted the following general inferences for the practitioner:

1. Function allocation cannot be accomplished by a formula—for example, rules that may apply in one situation may be irrelevant in another.

Table 6.4. Example of an Ergonomics Checklist for Equipment Operation

Characteristic	Satisfactory	Compromise but acceptable	Unsatisfactory
Console shape/size			
Desk height, area, control reach			
Display view			
Body, limb clearance			
Panel/location			
Frequency of use			
Sequence of use			
Emergency response			
Multioperator use			
Panel layout			
Functional grouping			
Sequential organization			
Identification			
Clearance spacing			
Displays			
Functional compatibility for intended purposes			
Intelligibility information content			
Control interaction			
Legibility; figures, pointers, scales			
Visibility; illumination, parallax			
Location			
Identification			
Controls			
Functional compatibility for intended purposes			
Location, motion excursion, and force			
Display interaction			
Spacing, clearance, size			
Identification			

Adapted from Woodson (1981).

Table 6.5. Checklist for Situation Analysis

What are the likely situations that could arise during system use, and how will these affect use of the system?	
Equipment	Falls short of target performance
	Falls short of specification
	Fails
Availability	Data are missing
	Materials are missing
	Personnel are missing
	Support is missing
Overloads	Of people/machines
	Of data, information, materials, and so on
Interruptions	The process breaks down
	Complete restart of process required
Environment	Changes in physical or social environment
Policy changes	Changes in laws, rules, standards, and guidelines

Adapted from Booth (1989).

Table 6.6. Fitt's List

Humans appear to surpass present-day machines with respect to the following:

- Ability to detect small amounts of visual or acoustic energy
- Ability to perceive patterns of light or sound
- Ability to improvise and use flexible procedures
- Ability to store very large amounts of information for long periods of time and to recall relevant facts at the appropriate time
- Ability to reason inductively
- Ability to exercise judgment

Present-day machines appear to surpass humans with respect to the following:

- Ability to respond quickly to control signals and to apply great force smoothly and precisely
- Ability to perform repetitive, routine tasks
- Ability to store information briefly and then erase it completely
- Ability to reason inductively, including computational ability
- Ability to handle complex operations, in other words, to do many different things at once

Adapted from Fitts (1951).

2. Function allocation is not a one-shot decision. The final assignment depends on the activities at the levels of the tasks, the conflation of tasks into jobs, the relationships of jobs in a larger workgroup, and the likely changes in the higher level manufacturing process.
3. Function allocations can be systemized—it is clear that there are number of sequential steps that can be taken to best allocate functions.
4. Both humans and machines can be good or bad at certain tasks.
5. Using analogies can facilitate the function allocation process.
6. Function allocation can be targeted to a specific time frame.
7. Function allocation depends on the nature of the task—it varies based on whether the task is perceptual, cognitive, or psychomotor.
8. Function allocation decisions must be based on sound economic analyses of options as well as the capabilities and limitations of the humans and machines.
9. Human and machine performances are not always antithetical.
10. Function allocation decisions must consider technology advances within a given time frame.
11. In cases where both humans and machines can perform a function, the system should be designed in such a way so that humans can delegate the function to machines or can take over the function when circumstances demand it.

A number of approaches have been suggested in the literature for solving the function allocation problem. Some of the promising approaches include function allocation criteria based on specific performance measures (e.g., time required to complete tasks; Mital, 1991, 1992; Mital & Mahajan, 1989; Mital, Mahajan, & Brown, 1988; Paul & Nof, 1979); comparison of capabilities and limitations of humans with particular attention given to knowledge, skills, and information sources and channels (Anderson, Back, & Wirstad, 1979; Badaracco, 1990; Brainbridge & Quintanilla, 1989; Ehn, 1988; Engstrom, 1991; Genaidy & Gupta, 1992; Prahalad & Hamel, 1990; Rasmussen, 1980; Shipley, 1995; Wirstad, 1998); economics (e.g., allocating the function to the less expensive option; Mital & George, 1989; Mital, Mahajan, & Brown, 1988; Mital & Vinayagamoorthy, 1987); and safety (to the human operator in the system; Hartley, 1983; Jian & Cheng, 1992; Mital & George, 1988).

Experiments with these approaches suggest that functions that are well proceduralized, which permits algorithmic analysis and requires little creative input, are prime candidates for automaton. Functions requiring cognitive skills of a higher order, such as design, planning, monitoring, exception handling, and so on are functions that are better performed by humans. The prime requirements for automating any function are the availability of a model of the activities necessary for that function, the ability to quantify that model, and a clear understanding of the associated control and information requirements. Clearly, there are some functions that should be performed by machines because of the following:

1. Design accuracy and tolerance requirements
2. Humans incapable of performing activity
3. Speed and high production volume requirements
4. Size, force, weight, and volume requirements
5. Hazardous nature of the activity

Equally, there are some actions that should be performed by humans because of the following:

1. Information acquisition and decision-making needs
2. Higher-level skill needs such as programming
3. Specialized manipulation, dexterity, and sensing needs
4. Space limitations (e.g., work that must be done in narrow and confined spaces)
5. Situations involving poor equipment reliability or where equipment failure could prove catastrophic
6. Activities for which technology is lacking

Mital et al. (1994) provide a generic methodology in the form of decision-making flowcharts for the systematic allocation of functions between humans and machines (Figure 6.4). These flowcharts are based on the requirements of complex decision making, on a detailed safety analysis, and on a comprehensive economic analysis of the alternatives.

Information Input and Processing for Human–Machine Interaction Design

Reduced to a fundamental level, human interaction with automation can be said to be dependent on the information-processing ability of the human and on the exchange of information among the different elements in a system. Over the years, behavioral scientists have attempted to explain human information processing through various conceptual models and themes. One such theory is the information theory (Vincete, 1990).

Information, according to the information theory, is defined as the reductions of uncertainty. Implicit in this definition is the tenet that events that are highly certain to occur provide little information, whereas events that are highly unlikely to occur, on the other hand, provide more information. Rather than emphasize the importance of a message in defining information, the information theory considers the probability of occurrence of a certain event in determining if there is information worth considering.

Information is measured in bits (denoted by H). One bit is defined as the amount of information required to decide between two equally likely alternatives. When the different alternatives all have the same probability, the amount of information (H) is given by

$$H = \log_2 N \qquad (1),$$

where N is the number of alternatives. For example, when an event only has two alternatives associated with it and when the two alternatives are equally likely, the amount of information, in bits, is 1.0.

When the alternatives are not equally likely (i.e., the alternatives have different probabilities of occurrence), the information conveyed by an event is given by

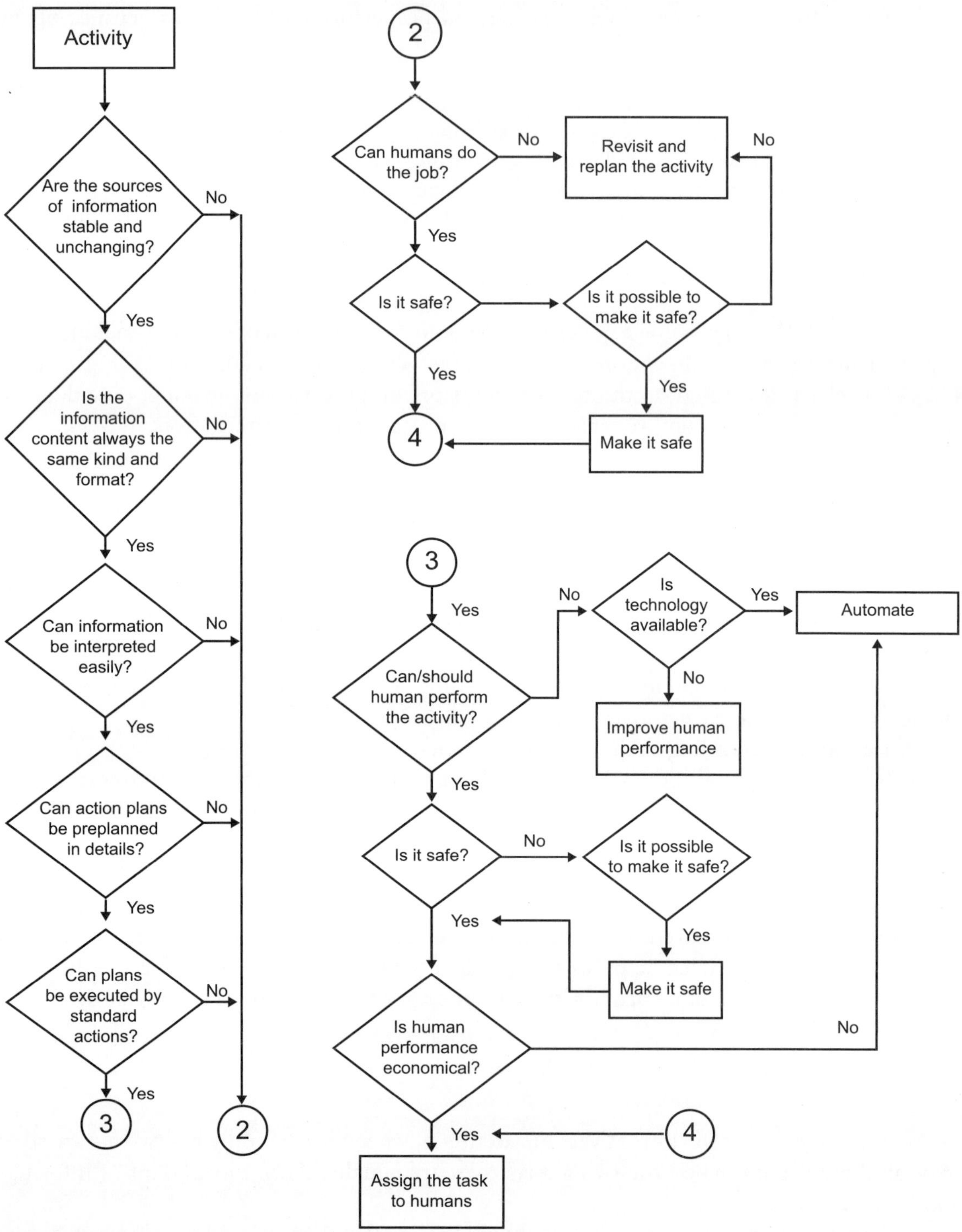

Figure 6.4. Generic function allocation analysis flowchart.

$$h_i = \log(1/p_i) \qquad (2),$$

where h_i is the information associated with event i, and p_i is the probability of occurrence of event i.

The average information (H_{av}) conveyed by a series of events having different probabilities is given by

$$H_{av} = \Sigma\, p_i\, (\log_2 (1/p_i)) \qquad (3),$$

where p_i is the probability of the event i.

Just as a bit is the amount of information, redundancy is the amount of reduction in information from the maximum due to the unequal probabilities of occurrence of events. Redundancy is expressed as a percentage, and is given by

$$\%\ \text{Redundancy} = (1 - H_{av} / H_{max}) \times 100 \qquad (4).$$

The channel capacity theory, another theory explaining information uptake by humans, is based on the premise that human sense organs deliver a certain quantity of information to the input end of a channel and that the output from the channel depends upon the capacity of the channel. It has been determined that if the input is small there is very little absorption of it by the channel, but if the input rises, it reaches the threshold channel capacity, beyond which the output from the channel is no longer a linear function of the input (Grandjean, 1988). Experimental investigations have shown that humans have a large channel capacity for information conveyed to them through the spoken word than through any other medium. A vocabulary of 2,500 words requires a channel capacity of 34 to 42 bits per second (Pierce & Karlin, 1957). Designers must keep in mind that in this day and age of information technology, the central nervous system of a human is subjected to more information than the information channel can handle and that a considerable reduction in the amount of information must be carried out before the human processes the information.

In addition to theories such as the information theory and the channel capacity theory, which explain information uptake, many conceptual models of human information processing have been proposed by researchers over the last four decades. Figure 6.5 shows one such fundamental model (most other models contain elements of this basic model) depicting the stages involved in information processing (Wickens, 1984). The key elements of the model are perception, memory, decision making, attention, response execution, and feedback. The following is a brief discussion of each of these elements.

Perception may involve detection (determining whether or not a signal is present) or identification and detection (involving detection and classification). The theory of signal detection (Green & Swets, 1988; Swets, 1988; Wickens, 1984) through the concept of noise in signals attempts to explain the process of perception and response to the perceived signals.

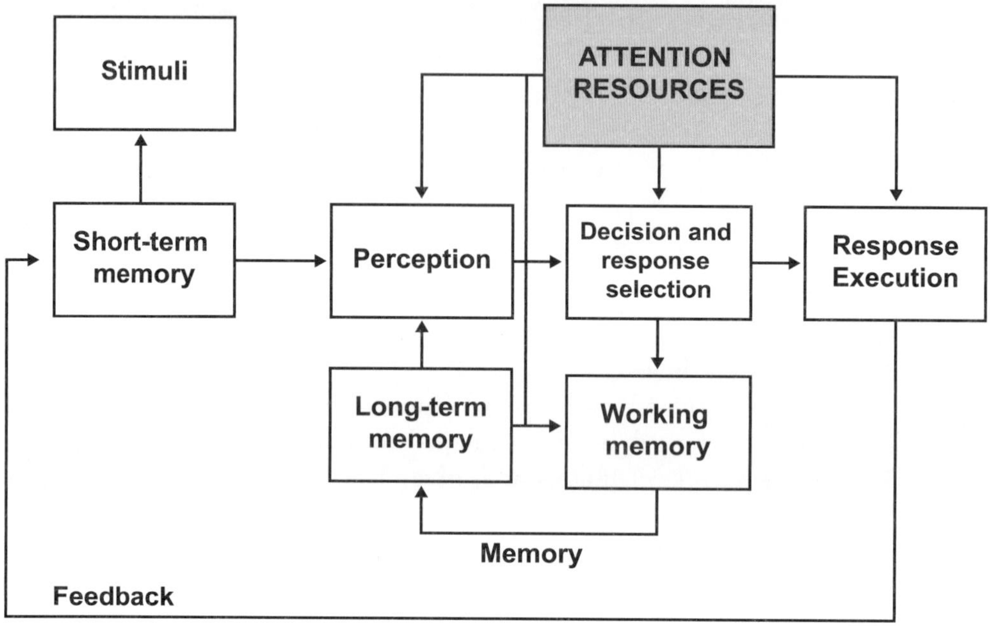

Figure 6.5. Fundamental model of human information processing.

Four possible outcomes are recognized in signal detection theory: (a) a hit (correctly concluding that there is a signal when there is one), (b) a false alarm (concluding that there is a signal when, in actuality, there is none), (c) a miss (concluding that there is no signal when, in actuality, there is one), and (d) a correction rejection (correctly concluding that there is no signal when there is none). The fundamental postulate of signal detection theory is that humans tend to make decisions based on criteria whose probabilities depend upon the probabilities of these outcomes. The probability of observing a signal, and the costs and benefits associated with these four possible outcomes, determine the responses of the human to the signal.

Memory, in humans, has been conceptualized as consisting of three processes, namely, sensory storage, working memory, and long-term memory (Wickens, 1984). According to this conception, information from sensory storage must pass through the working memory before it can be stored in long-term memory. Sensory storage refers to the short-term memory of the stimulus. Two types of short-term memory storage are well known—iconic storage associated with visual senses and echoic storage associated with the auditory senses (Sanders & McCormick, 1993). Sensory storage or short-term memory has been shown to be nearly automatic, requiring no sustained attention on the part of the human to retain it. An information transfer from sensory storage to working memory is brought about through attention (to the process). Information from stimuli is believed to be stored in the working memory primarily in the form of visual, phonetic, or semantic codes. It is also believed that the capacity of working memory is five to nine chunks of information (similar units regardless of the size; Miller, 1956). Researchers recommend presenting five to nine meaningful and distinct chunks of information for improved recall. It has also been determined that there is a linear relationship between the number of items in a memorized list and the time required to search the list of items in the working memory (Sternberg, 1966). Also, all items in the working memory are searched one at a time, even if a match is found early in the search process.

Rational decision making is defined as the process that involves seeking information relevant to the decision at hand, estimating the probabilities of various alternatives, and attaching values to the anticipated alternatives. A number of biases, however, have been identified to exist among humans that often make decision making irrational. Table 6.7 lists some of these biases.

Attention is another key factor influencing human information input and processing. Research has identified four types of tasks or situations requiring attention. These are selective attention, focused attention, divided attention, and sustained attention. When several information sources are to be monitored to perform a single task, attention is said to be selective (e.g., a process control operator scanning several instrument panels before detecting a deviant value). Table 6.8 provides guidelines for improving performances in tasks requiring selective attention.

When a human has to focus attention on one source of information and exclude all other sources of information for the task performance, attention is focused. The task performance

Table 6.7. Common Human Biases
Humans attach more importance to early information than subsequent information.
Humans generally do not optimally extract information from sources.
Humans do not optimally assess subjective odds of alternative scenarios.
Humans have a tendency to become more confident in their decisions with more information but do not necessarily become more accurate.
Humans tend to seek more information than they can absorb.
Humans generally treat all information as equally reliable.
Humans seem to have a limited ability to evaluate a maximum of more than three or four hypotheses at a time.
Humans tend to focus only on a few critical factors at a time and consider only a few possible choices related to these critical factors.
Humans tend to seek information that confirms their choice of action rather than information that contradicts or disconfirms their action.
Humans view a potential loss more seriously than a potential gain.
Humans tend to believe that mildly positive outcomes are more likely than mildly negative or highly positive outcomes.
Humans tend to believe that highly negative outcomes are less likely than mildly negative outcomes.

Adapted from Wickens (1984).

Table 6.8. Recommendations for Designing Tasks Requiring Selective Attention
Use as few signal channels as possible, even if it means increasing the signal rate per channel.
Inform the human about the relative importance of various channels for effective direction of attention.
Reduce stress levels on workers so more channels can be monitored.
Inform the human beforehand where signals will occur in the future.
Train the human to develop optimal scan patterns.
Reduce scanning requirements on the human by putting multiple visual information sources close to each other and by making sure the multiple sources of auditory information do not mask each other.
Provide a signal for a sufficient length of time for the individual to respond; where possible, provide for human control of the signal rate.
Adapted from Sanders and McCormick (1993).

under focused attention is affected by the physical proximity of the sources of information. While physical proximity enhances performance in tasks requiring selective attention, it impedes performance in tasks requiring focused attention. Table 6.9 provides guidelines for improving performance in tasks requiring focused attention.

When humans do more than one task at a time, their attention is divided (among the different tasks). While much of the theoretical base for explaining performance of tasks requiring divided attention is still evolving (Lane, 1982; Wickens, 1984), some guidelines for designing tasks that require divided attention are available and are provided in Table 6.10.

When humans maintain attention and remain vigilant to external stimuli over prolonged periods of time, attention is said to be sustained. Nearly 40 years of research in vigilance and vigilance decrement (Davies & Parasuraman, 1982; Graig, 1985; Parsuraman, 1986; Warm, 1984) has provided guidelines for improving performance in tasks requiring sustained attention (Table 6.11).

Table 6.9. Recommendations for Designing Tasks Requiring Focused Attention
Make the different channels of information as distinct as possible from the channel to which the human must attend.
Physically separate the channel of interest from the other channels.
Reduce the number of competing channels.
Make the channel of interest prominent by making it larger in size, or brighter, or louder, or by locating it centrally.
Adapted from Sanders and McCormick (1993).

Table 6.10. Recommendations for Designing Tasks Requiring Divided Attention

Minimize the potential sources of information.
Provide humans with a relative priority of tasks to optimize the strategy of divided attention.
Keep the level of difficulty of the task low.
Make tasks as dissimilar as possible in terms of task demands on the human.

Adapted from Sanders and McCormick (1993).

Table 6.11. Recommendations for Designing Tasks Requiring Sustained Attention

Provide appropriate work–rest schedules.
Provide task variation by interpolating different activities.
Make the signal larger, more intense, longer in duration, and/or distinct.
Reduce uncertainty in time and place of occurrence of the signal.
Use artificial signals and provide feedback to humans on their performance.
Reduce the rate of presentation of stimuli if it is high.
Provide optimal environmental conditions such as lighting, noise level, and so on.
Provide adequate training to humans to clarify the nature of signals to be identified.

Design Issues in Human Control of Systems

Humans respond to information and take controlling actions. The controlling actions of the human are mediated through the motor system in the human body. The human skeletal system, the muscles, and the nervous system help bring into play motor skills that enable the human to respond to stimuli. A motor skill is defined as the "ability to use the correct muscles with the exact force necessary to perform the desired response with proper sequence and timing" (Jensen, Schultz, & Bangerter, 1983). In addition, skilled performance requires adjusting to changing environmental conditions and acting consistently from situation to situation (Kelso, 1982).

A number of different types of human movements have been recognized by Sanders and McCormick (1993). These include discrete movements (involving a single reaching movement to a target that is stationary), repetitive movements (a single movement is repeated), sequential movements (a number of discrete movements to stationary targets), continuous movements (involving muscular control adjustments during movement), and static positioning (maintaining a specific position of a body member for a specified period of time). In addition, certain theoretical models of human motor responses explain the control aspects of human responses based on only two fundamental types of movements—fast and slow. Closed-loop theories (Adams, 1976, 1981), whether the movement is fast or slow, use the concept of sensory feedback (sensory information available during or after the motor response) to explain motor responses (to correct or reduce errors obtained through feedback).

The sensory receptors for feedback and feed forward (sensory information available prior to the action that regulates and triggers responses) are believed to be located in the muscle spindles (for sensing the muscle length and the rate of change of length) and tendons (Kelso, 1982; Winstein & Schmidt, 1989). The open-loop theories, on the other hand, are based on the belief that there are motor programs structured at a higher level containing information necessary for patterning the different movements (Grillner, 1985; Klapp, Anderson, & Berrian, 1973). Different deficiencies, such as the error of selection (where a person calls the wrong motor program for a controlling action) and the error of execution (where the correct motor program fails during execution of controlling actions) have been identified with motor programs (Schmidt, 1982).

The time it takes for the human to respond to stimuli is another critical factor that has been studied extensively in the literature (Sanders & McCormick, 1993). An understanding of the response time of the human is essential for the good design of the tasks involved in human interaction with automated systems. Response time is, in general, composed of reaction time and movement time. Reaction time is defined as the time from the signal onset calling for a response to the beginning of the response. Simple reaction time (reaction time in the presence of a single source of stimulus) has been shown to be between 0.15 seconds and 0.20 seconds. The mode through which the single stimulus occurs (visual, auditory, etc.), the detectibility of the stimulus (intensity, duration, and size), the frequency of the stimulus, the preparedness of the human for the stimulus, and the location of the stimulus (e.g., location in the peripheral field of view) are among the factors that have been shown to affect simple reaction time. Choice reaction time (reaction time in the presence of one of several possible stimuli each with different possible responses) is a function of the probability of a stimulus occurring; in other words, the reaction time is faster for events with greater probability. It has been shown to increase by about 0.15 seconds for each doubling of the number of possible alternative stimuli (Woodworth, 1938). Choice reaction time has been shown to be influenced by numerous factors, including the degree of compatibility between stimuli and responses, practice, the presence or absence of a warning signal, the type and complexity of the movement involved in the responses, and whether or not more than one stimulus is present in the signal. It has been determined that movement time is a logarithmic function of the distance of movement when the target size is a constant, and further that movement time is a logarithmic function of target size when the distance of movement is constant. This finding is popularly known as Fitts's law (1954) and is represented as

$$MT = a + b \, Log_2(2D/W) \qquad (5),$$

where MT is the movement time, a and b are empirical constants dependent upon the type of movement, D is the distance of movement from start to the center of the target, and W is the width of the target.

Human response to stimuli is dependent not only on the speed of the response but also on the accuracy of the response. The accuracy of the human response assumes special importance when the response has to be made in situations where there is no visual feedback (a situation referred to as "blind positioning"). Movements that take place in a blind positioning situation have been determined to be more accurate when the target is located dead ahead than when located on the sides. Also, targets below shoulder height and the waist level are more readily reachable than targets located above the shoulder or the head (Fitts, 1947). The distance and speed of movement have also been found to influence the accuracy of the response (Brown, Knauft, & Rosenbaum, 1947; Schmidt, Zelaznik, Hawkins, Frank, & Quinn, Jr., 1979).

Display Design for a Human–Machine Interaction

A display is defined as any indirect means of presenting information. Displays are generally one of the following four types: visual, auditory, tactual, and olfactory. The visual and the auditory modes of displaying information are the most common. Displays based on tactile and olfactory senses are mostly used for special tasks or user situations (e.g., for the hearing impaired).

Selecting the mode of display, whether it should be visual or auditory in nature, is an important factor due to the relative advantages and disadvantages certain modes of display may have over other modes for specific types of task situations (the auditor mode is better than visual displays in vigilance), environments (lighting conditions), and user characteristics (a person' s information-handling capacity). Table 6.12 provides general guidelines for deciding between two common modes of information presentation, namely, auditory and visual.

The types of displays to present information also depend on the type of information to present. Different types of information can be presented using displays when the sensing mode is indirect. Information can be either dynamic or static. Dynamic information is categorized by changes occurring in time (e.g., fuel gage). Static information, on the other hand, does not change with time (e.g., printed safety signs). A number of other types of information are also recognized in the literature.

Control Design for Human–Machine Interaction

Controls are the primary means of transmitting the controlling action to devices and systems. Numerous factors affect the design of control devices. These factors include the ease of identification, the size of the control, lag, backlash, dead space, and location.

The ease of identification of controls depends upon how well the controls have been coded. The efficacy of the coding used can be determined using detectability, discriminability, compatibility, meaningfulness, and the extent of standardization of the codes. Controls can be coded using shape, texture, size, location, operational methods, colors, and labels. Shape coding uses tactual sensitivity of the human for discriminating between the different

Table 6.12. Guidelines for Deciding When to Use Visual Displays and When to Use Auditory Displays

Characteristics		*Visual displays*	*Auditory displays*
Message characteristics			
Simple message			✓
Complex message		✓	
Short message			✓
Long message		✓	
Potential reference value of message	High	✓	
	Low		✓
Immediacy of action requirement of message	High		✓
	Low	✓	
Message deals with events in time			✓
Message deals with location in space		✓	
Human capability			
Auditory system overburdened		✓	
Visual system overburdened			✓
Environmental factors			
Location to bright or too dark requiring significant adaptation			✓
Location too noisy		✓	

Adapted from Deatherage (1972).

shapes of the controls. Figure 6.6 provides examples of different shapes that are commonly used in controls. Three different types of textures have been identified as being suitable for coding control devices: smooth surface, fluted surface, and knurled surface.

The most important consideration when coding by size is to provide adequate discriminability between the different sizes used. There are instances when coding is based on the method of operation of the control (e.g., push button controls). Such operational coding is undesirable, however, when operation time or potential operator errors are considerations. Another way to code controls is by color. Meaningful colors (such as red for a danger button), combined with other coding dimensions such as shape and size, have been shown to be effective in enhancing the discriminability of the controls. A common method of coding controls is by labels. Labels are even considered a minimum requirement in many situations as they do not place extensive learning demands on the operators. Labels, however, have the disadvantage in that they take time to read and are not useful as a coding method in situations that have a high operation speed. Also, the placement of the label on the control has

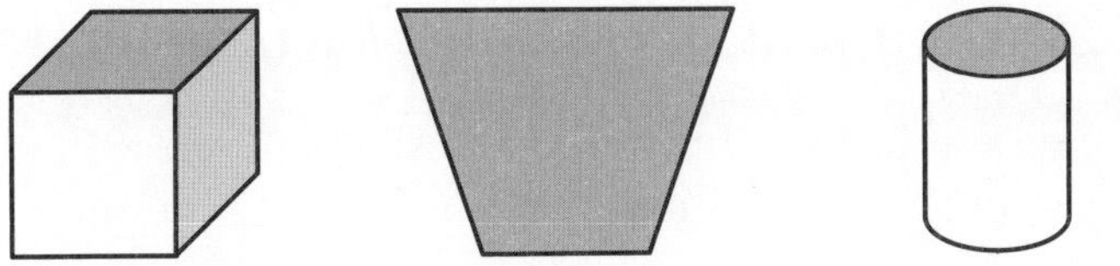

Figure 6.6. Coding control devices using shapes.

been shown to pose accessibility problems to the reader. Control devices can have unique combinations of codes or even redundant codes. Considerations such as the illumination of the panel, the potential visual handicaps of the operator, the maintenance of mechanical controls, and the speed and the accuracy with which the controls have to be operated are other factors to consider in designing controls for ease of identification.

The control response (C/R) ratio is an important control design consideration. A sensitive control will have a low C/R ratio (i.e., the response will be large even for a slight change in the control). It is believed that human motor action is at two levels—at a gross adjustment level and at a fine adjustment level. Hence the optimal level of C/R ratio to use in a control device is generally decided as a trade-off between the time it takes to accomplish the gross movement and the time it takes for the fine adjustment involved in a controlling action. It has been shown that an optimum C/R ratio is dependent upon certain equipment- and design-related factors, such as the type of control (lever, crank, wheel, etc.), the size of the display, and the tolerance permitted in setting the control.

Resistance in a control is responsible for providing feedback about the controlling action to the operator. Control resistance is made up of two fundamental elements: the force applied to the control and the distance to which this force is applied (or the distance to which the control moves). Most controls use a combination of both pure displacement and pure force mechanisms for providing operator feedback. Control resistance can significantly affect operator performance by affecting the speed and precision of control operations, by changing the feel in the control, by changing the smoothness of the control movement, and by subjecting the control to the effect of shock and vibration.

Dead space is defined as the amount of movement near the null position of the control. The amount of dead space in a control device has been shown to affect the sensitivity, and hence the performance, of the control system. It has been shown by researchers that dead space in a control device can be compensated to a certain extent by making the control response ratio less sensitive.

Backlash in a control device is defined as the dead space at any control position. Research on backlash shows those systems with high control gain need to have minimum backlash to reduce system errors. If the control system design makes it impossible to reduce the backlash, the recommendation is to make the control gain as low as possible, since humans have shown to cope with backlash errors poorly.

Other Design Considerations in Human–Machine Interaction Design

Besides the individual design factors affecting the design and operation of displays and controls, compatibility, grouping, and location of controls impact design of human interaction with technology.

Compatibility is the relationship between the expectations of the human and the input stimuli and responses of the system with which the human is interacting. Any system with human users should be compatible with the human expectations. In general, good compatibility will result in fewer user errors and better human and overall system performance. Literature identifies four types of compatibility: conceptual, movement, spatial, and modality compatibilities (Miller, 1956).

Conceptual compatibility refers to the matching that should exist between certain forms of stimuli such as symbols and the conceptual associations humans make with such stimuli. Movement compatibility (also commonly referred to as population stereotypes) denotes the relationship between the movement of the displays and controls and the output response of the system being controlled. The most important types of movement compatibilities include the movement of a control to follow the movement of a display, the movement of a control to control the movement of a display, the movement of a control to produce a specific system response, and the movement of a display without any related response. Spatial compatibility refers to the relationship that should exist between the physical features and arrangement of the controls and their associated displays. A good example of compatibility in the physical features between the displays and the controls is the design of the function keys on a keyboard and the corresponding labels for the function keys. In a number of experiments with household stovetops, human factor researchers have demonstrated conclusively the need for physically arranging displays and the associated controls in a corresponding and compatible way. Modality compatibility is a fairly new addition to the list and refers to certain stimulus-response combinations being more compatible with some tasks than with others.

The control-display arrangement in a workspace has to be based on the human sensory capabilities and the anthropometric, biomechanical, and other characteristics of the human user. A few general principles of control-display location and arrangement are useful in setting priorities and in determining trade-offs for good design, if not optimal. According to the importance principle, components that are vital to system goals should be placed in convenient locations. System experts determine what these vital goals are. According to the frequency-of-use principle, components that are frequently used should be placed in convenient locations. According to the functions principle, components that are functionally related in the operation of the overall system should be grouped and placed together. According to the sequence-of-use principle, components should be arranged in the sequence in which they find frequent use in the operation of the system or in the performance of a task. Use of one or a combination of these principles requires that the system designer collect information about the human users involved, the tasks involved, and the environment in which the user has to perform the task.

Summary

This chapter presented an overall process for designing and evaluating human interaction with technologies in a modern-day production or service system. The key elements involved in this process were briefly described. Humans and technologies are essential and valuable resources in the industrial workplace. Humans and technologies have a symbiotic relationship with each other, and one cannot function without the other. The challenge, then, is to jointly optimize human resources and technology resources to maximize productivity and effectiveness of operations. The chapter discussed the scientific bases behind designing the human technology interaction in the industrial workplace, identified key design challenges, and reviewed tools and techniques to overcome the design challenges.

REFERENCES

Adams, J. (1976). Issues for a closed-loop theory of motor learning. In G. Stelmach (Ed.), *Motor control: Issues and trends* (pp. 87–101). New York, NY: Academic Press.

Adams, J. (1981). A closed-loop theory of motor learning. *Journal of Motor Behavior*, *3*, 111–150.

Anderson, H., Back, P., & Wirstad, J. (1979). *Job analysis for training design and evaluation—description of a job analysis method for process industries (Report No. 6)*. Karlstad, Sweden: Ergonomrad.

Badaracco, J. (1990). *The knowledge link*. Cambridge, MA: Harvard Business School Press.

Blackman, H. S., Gertman, D. I., & Gilmore, W. E. (1983). *CRT display evaluation: The checklist evaluation of CRT—generated displays* (Report No. NUREG/CR-3557). Washington, DC: Nuclear Regulatory Commission.

Booth, P. A. (1989). *An introduction to human–computer interaction*. Hove, England: Lawrence Erlbaum Associates.

Brainbridge, L. S., & Quintanilla, S. A. R. (1989). *Developing skills with information technology*. Chichester, England: John Wiley & Sons.

Brown, J., Knauft, E., & Rosenbaum, G. (1947). The accuracy of positioning reactions as a function of their direction and extent. *American Journal of Psychology*, *61*, 167–182.

Bulinger, H. J., & Warnecke, H. J. (1985). *Factory of the future*. Heidelberg, Germany: Springer-Verlag.

Cakir, A., Hart, D. J., & Stewart, T. F. M. (1980). *Visual units*. Chichester, England: John Wiley & Sons.

Chapanis, A. (1976). *Research techniques in human engineering*. Baltimore, MD: John Hopkins.

Davies, D., & Parasuraman, R. (1982). *The psychology of vigilance*. London, England: Academic Press.

Deatherage, B. H. (1972). Auditory and other sensory forms of information presentation. In H. P. Van Cott & R. Kinkade (Eds.), *Human engineering guide to equipment design*. Washington, DC: U.S. Government Printing Office.

Department of Defense. (1981). *Human engineering criteria for military systems, equipments and facilities* (Publication No. MIL-STD 1472C). Washington, DC: U.S. Department of Labor.

Ehn, P. (1988). *The work-oriented design of computer artifacts*. Stockholm, Sweden: Almquist and Wiksell International.

Engstrom, T. (1991). Future assembly work–natural grouping. In *Designing for everyone—Proceedings of the XIth Congress of the International Ergonomics Association* (Vol. 2, pp. 1317–1319). London, England: Taylor & Francis.

Fitts, P. (1947). A study of location discrimination ability. In P. Fitts (Ed.), *Psychological research on equipment design* (Research report 19). Aremy, Air Force, Aviation Psychology Program, Ohio State University, Columbus, OH.

Fitts, P. (1951). *Human engineering for an effective air navigation and traffic-control system*. Washington, DC: National Research Council.

Fitts, P. (1954). The information capacity of the human motor system in controlling the amplitude of movement. *Journal of Experimental Psychology*, *47*, 381–391.

Genaidy, A. M., & Gupta, T. (1992). Robot and human performance evaluation. In M. Rahimi & W. Karwowski (Eds.), *Human–robot interaction* (pp. 4–15). London, England: Taylor & Francis.

Graig, A. (1985). *Vigilance: Theories and laboratory studies*. In S. Folkard & T. Monk (Eds.), *Hours of work: Temporal factors in work-scheduling* (pp. 107–121). Chichester, England: Wiley.

Grandjean, E. (1988). *Fitting the task to the man* (4th ed.). London, England: Taylor & Francis.

Green, D., & Swets, J. (1988). *Signal detection theory and psychophysics*. Los Altos, CA: Peninsula.

Grillner, S. (1985). Neurobiological bases of rhythmic motor acts in vertebrates. *Science*, *228*, 143–149.

Hartley, J. (1983). *Robots at work: A practical guide for engineers and managers*. Bedford, England: IFS.

Health and Safety Executive. (1989). *Human factors in industrial safety (health and safety guidance booklets)*. London, England: HSE Books.

Jensen, C., Schultz, G., & Bangerter, B. (1983). *Applied kinesiology and biomechanics*. New York, NY: McGraw-Hill.

Jiang, B. C., & Cheng, O. S. H. (1992). Six severity level design for robotic cell safety. In M. Rahimi & W. Karwowski (Eds.), *Human–robot interaction* (pp. 161–180). Washington, DC: Taylor and Francis.

Kelso, J. (1982). *Human motor behaviour: An introduction*. Hillsdale, NJ: Lawrence Erlbaum Associates.

Kirwan, B., & Ainsworth, L. K. (Eds.). (1992). *A guide to task analysis*. London, England: Taylor & Francis.

Klapp, S., Anderson, W., & Berrian, R. (1973). Implicit speech in reading, reconsidered. *Journal of Experimental Psychology*, *100*, 368–374.

Lane, D. (1982). Limited capacity, attention allocation, and productivity. In W. Howell & E. Fleishman (Ed.), *Human performance and productivity: Information processing and decision making*. Hillsdale, NJ: Lawrence Erlbaum Associates.

Meister, D. (1986). *Human factors testing and evaluation*. New York, NY: Elsevier Science Publishers.

Miller, G. (1956). The magical number seven, plus or minus two: Some limits on our capacity for processing information. *Psychological Review*, *63*, 81–97.

Mital, A. (1991). Manual versus flexible assembly: A cross-comparison of performance and cost in four different countries. In M. Pridham & C. O'Brien (Eds.), *Production research: Approaching the 21st century*. London, England: Taylor & Francis.

Mital, A. (1992). Economics of flexible assembly automation: Influence of production and market factors. In H. R. Parsaei & A. Mital (Eds.), *Economics of advanced manufacturing systems* (pp. 45–72). London, England: Chapman & Hall.

Mital, A. (1997). What role for humans in computer integrated manufacturing? *International Journal of Computer Integrated Manufacturing*, *10*, 190–198.

Mital, A., & George, L. J. (1988). Human issues in automated (hybrid) factories. In F. Aghazadeh (Ed.), *Trends in ergonomics/human factors V* (pp. 373–378). Amsterdam: North Holland.

Mital, A., & George, L. J. (1989). Economic feasibility of a product line assembly: A case study. *Engineering Economist*, *35*, 25–38.

Mital, A., & Mahajan, A. (1989). Impact of production volume, wage, and interest rates on economic decision making: The case of automated assembly. In *Proceedings of the Conference of the Society for Integrated Manufacturing* (pp. 558–563). Norcross, GA: Institute of Industrial Engineers.

Mital, A., Mahajan, A., & Brown, M. L. (1988). A comparison of manual and automated assembly methods. In *Proceedings of the IIE Integrated Systems Conference* (pp. 206–211). Norcross, GA: Institute of Industrial Engineers.

Mital, A., Morotwala, A., Kulkarni, M., Sinclair, M., & Siemieniuch, C. (1994). Allocation of functions to humans and machine in a manufacturing environment: Part II—the scientific basis (knowledge base) for the guide. *International Journal of Industrial Ergonomics*, *14*, 33–49.

Mital, A., & Pennathur, A. (2004). Advanced technologies and humans in manufacturing workplaces: An interdependent relationship. *International Journal of Industrial Ergonomics*, *33*, 295–313.

Mital, A., & Vinayagamoorthy, R. (1987). Case study: Economic feasibility of a robot installation. *Engineering Economist*, *32*, 173–196.

Nuclear Regulatory Commission. (1981). *Guidelines for control room design reviews* (Report No. NUREG-0700). Washington, DC: U.S. Nuclear Regulatory Commission.

Parsuraman, R. (1986). Vigilance, monitoring, and search. In K. Boff, L. Kaufmann, & J. Thomas (Eds.), *Handbook of perception and human performance: Cognitive process and performance*. New York, NY: Wiley.

Paul, R. P., & Nof, S. (1979). Work methods measurements—a comparison between robot and human task performance. *International Journal of Production Research*, *17*, 277–303.

Pierce, J. R., & Karlin, J. E. (1957). Reading rates and the information rate of a human channel. *Bell System Technical Journal*, *36*, 497–516.

Prahalad, C. K., & Hamel, G. (1990). The core competence of the corporation. *Harvard Business Review*, *68*, 79–91.

Rasmussen, J. (1980). *Some trends in man–machine interface design for industrial process plants (Report No. Riso-M-2228)*. Roskilde, Denmark: Riso National Laboratory.

Sanders, M. S., & McCormick, E. J. (1993). *Human factors in engineering and design*. New York, NY: McGraw-Hill.

Sarter, N. B., & Woods, D. D. (1997). Team play with a powerful and independent agent: Operational experiences and automation surprises on the airbus A-320. *Human Factors*, *39*, 553.

Schmidt, R. (1982). *Motor control and learning: A behavioral emphasis*. Champaign, IL: Human Kinetics.

Schmidt, R., Zelaznik, H., Hawkins, B., Frank, J., & Quinn, J., Jr. (1979). Motor output variability: A theory for the accuracy of rapid motor acts. *Psychological Review*, *86*, 415–451.

Seminara, J. (1949). *Human factors methods for assessing and enhancing power plant maintainability (Report No. EPRINP-2360)*. Palo Alto, CA: Electric Power Research Institute.

Sheridan, T. B. (1987). Supervisory control. In G. Salvendy (Ed.), *Handbook of human factors*. New York, NY: John Wiley & Sons.

Shipley, P. (1995). The analysis of organizations as an aid for ergonomics practice. In J. R. Wilson & E. N. Corlett (Eds.), *Evaluation of human work: A practical ergonomics methodology*. London, England: Taylor & Francis.

Sternberg, S. (1966). High-speed scanning in human memory. *Science*, *153*, 652–654.

Swets, J. (Ed.). (1988). *Signal detection and recognition by human observers: Contemporary readings*. Los Altos, CA: Peninsula.

Warm, J. (Ed.). (1984). *Sustaining attention in human performance*. Chichester, England: Wiley.

Wickens, C. (1984). *Engineering psychology and human performance*. Columbus, OH: Merrill.

Winstein, C., & Schmidt, R. (1989). Sensorimotor feedback. In H. Holding (Ed.), *Human skills* (2nd ed.). Chichester, England: Wiley.

Wirstad, J. (1998). On knowledge structures for process operators. In L. P. Goodstein, H. B. Anderson, & S. E. Olsen (Eds.), *Tasks, errors and mental models*. London, England: Taylor & Francis.

Woodson, W. E. (1981). *Human factors design handbook*. New York, NY: McGraw-Hill.

Woodson, W. E., Tillman, B., & Tillman, P. (1991). *Human factors design handbook: Information and guidelines for the design of sciences, facilities, equipment, and products for human use*. New York, NY: McGraw-Hill.

Woodworth, R. (1938). *Experimental psychology*. New York, NY: Henry Holt.

CHAPTER 7

WORK SYSTEM DESIGN USING ISO STANDARD AND ECOLOGICAL APPROACHES TO ENHANCING EFFECTIVENESS OF HUMAN RESOURCES MANAGEMENT

Yoshio T. Ikeda

Department of Business Administration
Aichi Institute of Technology

OVERVIEW

To find problems and issues in working systems and their environment, we referred to the official or public statistics such as labor white papers (or white books) and other private statistics issued in Japan. The identified data have suggested that a corporate recruiting schema has offered more jobs with limited term employment (especially for younger workers) than those with an unlimited term in the last two decades. It is also noted that senior workers have not been offered proper jobs. The limited term system and the senior unemployment tendency could increase the degree of freedom for younger employees to choose a job with their working style. But it brought unfair payment and treatment to them. Unemployment issues in term of senior workers, of course, created unfair treatment between the employer and employee as well.

To solve part of these problems, ISO 6385-2004 is reviewed and introduced as a basic framework for professionals and other people to (re)design work systems and working situations for improvement. The author attempted to utilize the ecological approaches to apply International Organization for Standardization (ISO) design requirements to human–machine systems, and he integrated human resource management and practice for a solution that could satisfy the employers and employees. What resulted was a formulation of the points that characterize the Ecological Approach to Human Machine Systems (EAHMS) and Integrated Human Resources Management (IHRM) method, which are as follows: (a) the bottom-up decision-making system for well-motivated workers, (b) the closed-loop feedback system between and among workers and managers, (c) jobs abstraction for intersectional work analysis and integrated management system, and (d) information and communication technology based on a knowledge-sharing system, especially a knowledge database system. These points with others provide managers (employers) and workers (employees) with well-being.

Having reviewed the problems of 20th-century human resource management, 21st-century human resource recruiting schema is discussed as requirements. Also discussed are ecological approaches to human–machine (environment) systems, to integrated human resource management, and to human resource development (organizational learning) as perspective.

It is proposed that successful integration of all the human resource management (HRM) and business systems related to the triad of recruiting schema (human resource recruitment), the ecological approach (human resource management), and organizational learning (human resource development) is coherently formed in languages for an IHRM operating platform.

INTRODUCTION

This chapter describes the analyzed working environment issues from the aspects of personnel management statistics. In 1973, the Organization for Economic Cooperation and Development (OECD) survey group investigated Japanese labor and personnel management systems. They reported that the businesses have been managed and operated under three characteristics of labor and personnel management systems, which include lifetime employment, the seniority system, and a labor union within a company. These were supposed to be strong components in the business operations because they could have overcome the economic recessions that occurred in the 1970s and 1980s. However, it seems that many companies have changed their corporate recruiting schema from lifetime employment to temporary employment in the last two decades. The tendency toward just temporary employment could actually increase the freedom of job choice for younger employees who want to choose. Senior workers are not kept gainfully employed. This results in unfair age discrimination too. Unemployment issues for senior workers, of course, also create friction between the employer and employee. These changes have brought negative effectiveness to human resource management.

Since the 1980s, the implementation of information technology and business globalization have been changing work conditions for workers and their social lifestyles. With the advent of information communication technology (ICT), workers interact more with computer-based interfaces to perform their jobs than directly with coworkers. Business globalization has made businesses move entire resources across national borders. Within an organization, humans are the only typical resource compared with the other resources of equipment, raw materials, capital, and energy because humans have minds and lives under sociocultural backgrounds, whereas other resources do not have them.

Moreover, humans play important roles in an organization as decision makers in planning and controlling and as actors in getting jobs done. Although globalization has brought a sort of efficiency to business operations, it also has created such problems of effectiveness as misunderstanding and miscommunications among domestic and international workers due to factors that include moral hazards, cultural gaps, social class differences based upon employment types, the level of computer literacy, and others. These shortcomings have affected workers in their work and home life, especially those who have poor social lives, the so-called losers in the society. The moral hazard is an intangible in its environment but an important factor for maintaining organizational effectiveness and efficiency. Therefore, there are two major factors behind human resource management: the ability to work with computer-based systems and work with mutual and cooperative systems, which would be more important in the long term for an organization to survive in a competitive market.

Therefore, this chapter starts with the results of the analyzed issues from the aspects of personnel management statistics that depict the problems of personnel management. In the second section, the ISO ergonomic work system design principles are reviewed and introduced as the basis of collaborative workplace design schema. The next section introduces the EAHMS and IHRM as a method that could satisfy the requirements of employers and employees (Ikeda, 2003). It can be used as a framework of bottom-up decision making, closed-loop feedback, and job satisfaction accommodated in job descriptions, resulting in both employers and employees forming harmonious working relationships. Finally, the chapter deals with the abstraction hierarchy with a shift of languages as a possible method of a bottom-up and closed-loop system implementation.

HRM Problems Indicated by Statistics of Personnel Management

National and international economic conditions affect many industries in their business performance. Since human resources are situated in work organizations, it is worthwhile to find HRM problems indicated by the statistics of personnel management. In many workplaces, the number of nonregular employees (e.g., part-time workers) is increasing in companies with more than 500 employees, as indicated in Table 7.1 (Ikeda & Masuzawa, 2009), The companies surveyed with more than 1,000 employees stated that the reasons why they increased the number of nonregular employees are the labor cost reduction (91.2%) and reduction of training (56.2%). The percentage of organized workers belonging to an in-house union decreased steeply to 18.1% in the year 2002. The ratio of part-time workers to regular workers in Japan is now exceeding those in the United States and France. These significant issues are outlined in Table 7.1.

The efficiency and/or effectiveness of business is based upon the accumulation of the individual effort of teamwork that should be synergetic (Ikeda, 2003). Therefore, it is important that team coordination and communication should be included in the design of a workplace. Recent labor statistics show, however, that personnel management practices tend to aim toward an ideal system. This trend is a major problem for HRM to maintain its effectiveness and efficiency. Surveys indicate that the attitude of workers showed more of a decreasing rate of loyalty to their respective companies than ever before. See Table 7.1 for a list of reasons explaining why employees are less loyal to their companies.

In conclusion, workers in different ranks in the company receive differential treatment. Workers at the lower ends of the organizational hierarchy do not feel comfortable expressing their opinions, causing employees to be stressed. Employees, therefore, become less motivated and loyal to their team or company. The management as well is frustrated and relies heavily on part-timers who are not well qualified. This causes many HRM problems and issues that should be answered immediately. The problems and issues are often affected by working environment factors, as reviewed in the following section.

HRM Problems Indicated by Articles of Working Environment Issues

Since human resources are situated in work organizations, it is also worthwhile to investigate HRM problems from the working environment point of view. The significant issues are stated in Table 7.2 (these are selected from over 1,000,000 articles issued by both Japanese and overseas reliable or influential media from 2007 to 2009; Masuzawa & Ikeda, 2009). Table 7.2 is supported by statistics issued by reliable officials or public statistics agencies shown in Table 7.1. The term "freeter," coined from "a *free* part-time work*er*," means a worker who wants to be free to choose his or her employment term and work hours. The Japanese Internal Affairs and Communications Ministry states that part-time workers range from age 15 through age 34, not including students and married women.

Employees, therefore, become less loyal and less motivated. On the other hand, management is also frustrated so it relies heavily on temporary workers who are not well qualified. This trend causes many problems and issues in the work environment, which can be summarized as recruiting schema problems.

Table 7.1. HRM Problems Associated With Personnel Management

No.	*Questionnaire*	*Subjects*	*Data*	*Source*
1	How many part-timers are working?	Company	In 2007, nonregular employees (e.g., part-time workers) constituted 30.1% of total employees in companies (more than 500 employees), increased by 6.6% comparing with year 2002.	SB1
2	Why did you company increase the number of part-timers?	Company	In 2006, companies with more than 1,000 employees stated that the reason they increased the number of nonregular employees were cost reduction (91.2%), reduction of training (56.2%), and benefit of ICT and manuals (41.7%).	SL1
3	What kinds of skills are required for employees?	Company	2002 / 2007 / Dif. Rationalization skill 16.4 / 28.5 / +12.1 Line job skill (general) 30.5 / 19.4 / –11.1 Line job skill (assembly) 11.5 / 5.7 / –5.8 Equip. maintenance skill 9.3 / 12.0 / +2.7 QA and testing skill 5.9 / 10.7 / +4.8	SL2
4	What is the percentage of organized workers by in-house union?	Company	Decreased steeply from 35.6% in 1955 to 18.1% in 2002	MHLW1
5	Part-time worker/regular worker ratio	Company	Japan 17.6%; OECD 16.1%; France 13.3%; United States 12.6%	OECD1
6	Years of service (average)	Company	Japan 11.3; France 10.7; Germany 9.7; UK 7.8; USA 7.4	OECD1
7	Loyalty to the company	Worker	Higher than 3 years ago (16.2%), less (26.1%), and neither (51.7%).	SL3
8	Reasons for disloyal workers (unsatisfied with jobs)	Worker	Not well paid (46.8%), work assessment problem (35.6%), communication problem (24.2%), training and learning problem (21.1%), and promotion problem (21.4%)	SL3
9	Reasons for less incentive	Worker	Not well paid (43.4%), training and learning (career development) problem (32.6%), self-realization and esteem problem (28.9%), and social contribution problem (20.3%)	SL3

Table 7.1. (continued)

No.	*Questionnaire*	*Subjects*	*Data*	*Source*
10	Reasons of (first) resignation	Worker	Male: tedium of work (35.8%), communication problem (20.9%), wage problem (19.3%); Female: 20.4%, 16.1%, and 11.05, respectively	CO1
11	Which do you prefer, free time or high salary?	Worker	Time: 31.4% (1997) to 37.7% (2007); high salary: 49.2% (1997) to 44.7% (2007).	CO2
12	Which do you prefer, mental richness or physical richness?	Worker	Mental: 56.3% (1997) to 62.6% (2007); physical: 30.1% (1997) to 28.6 (200)	CO2

Source Indices (on all the companies and workers registered in Japan unless otherwise noted):
SB1: Japanese Statistics Bureau (2007).
SL1: Institute for Science of Labour (2006).
SL2: Institute for Science of Labour (2008b).
SL3: Institute for Science of Labour (2008a).
MHLW1: Japanese Ministry of Health, Labour, and Welfare (2008).
OECD1: OECD (2007).
CO1: Cabinet Office of the Japanese Government (2005).
CO2: Cabinet Office of the Japanese Government (2008).

Table 7.2. HRM Problems Associated With Working Environment

No.	*Article title (points of contents)*	*Summary of the articles*	*Related statistics no. in Table 7.1.*	*Source*
1	*In Japan, secure jobs have a cost* (cost reduction and nonregular workers)	The lifetime employment system, cemented in Japan's postwar economic boom, bound dutiful workers and paternalistic employers together, producing a mutual loyalty (and labor harmony) rarely seen in the West. But Japan's obsession with keeping workers employed—even those who are not needed—comes at a cost. Since Japanese workers are hard to lay off, companies have turned to temporary workers who receive lower pay and fewer benefits and can be cut more easily. Such workers now make up a third of Japan's work force. But keeping all those permanent staff members occupied is its own challenge.	2	*New York Times*, May 19, 2009
2	*Companies turn to new tactics to nurture fresh recruits* (countermeasures to high turnover of new employees)	Nurturing young employees is a major problem for some companies. They hold seminars and training courses in April for their new workers who graduated in March. In recent years, it has become much less rare for employees to quit just a few years after they are hired. Some firms try to teach their employees teamwork against the background of growing calls to deepen their understanding of and blend into the companies they work for. Three cases are as follows:	10	*Kyodo News*, Sept. 10, 2008

3	*Where is the "Japanese Dream"?* (failure to nurture younger workers)	Judging by the growing numbers of "freeters"—those underpaid and despised younger workers who basically run every aspect of Japan's economic machine and whose only dream is to save enough money to leave Japan and try their luck in the United States or Canada—it can be said that Japan's problem is mostly due to its failure to nurture and invest in people. Only the privileged get to experience the fruit of the country's mighty economic power. For the common people, Japanese and foreigners alike, Japan offers very little hope.	–	*Japan Times*, Mar. 9, 2008
4	*"Freeter" ranks ease but not woes* (freeter problem is harder for higher ages)	The number of "freeters"—nonstudent part-timers excluding housewives—continues to decline, but it is becoming harder for them to find stable work once they reach a certain age, according to a government white paper. The number of freeters aged 15 to 34 totaled 1.81 million in 2007, declining for the 4th year in a row and down from a peak of 2.17 million in 2003, according to the 2008 white paper on juveniles. Freeters aged 15 to 24 have declined by 300,000, or 25.2% since 2003. But freeters aged 25 to 34 fell by just 60,000, or 6.1%, over the same period, indicating it is harder for them to find permanent employment.	1, 5, 6	*Kyodo News*, Nov. 22, 2008
5	*Activist views homeless in realistic light* (the spiral of poverty)	Many temporary or contract workers tend to look for jobs that will also provide housing because they usually do not have enough savings to pay both the "key money" and deposit typically needed to rent an apartment. And because their pay level is too low to save anyway, it is difficult for them to get regular jobs and survive for a month or so needed to get their first paycheck. This vicious cycle has created even tougher labor conditions, generating more part-time workers while forcing regular employees to do more overtime. It's important to relate the issue of poverty to the issues of long working hours, "karoshi" (death due to overwork), and unemployment.	–	*Japan Times*, March 21, 2009

Table 7.2. (continued)

No.	Article title (points of contents)	Summary of the articles	Related statistics no. in Table 7.1.	Source
		1. Mary Chocolate Co. holds a "top seminar" for university students every year as a means of retaining them once they are hired in order to avoid high turnover. A Mary Chocolate official said the purpose of the seminar is to instill in the new recruits "creativity and careful attention" to others, adding that their efforts to follow through with the seminar "strengthen a sense of solidarity and as a consequence lead to a rise in the rate of new employees remaining with the company." 2. Benesse Corp., which specializes in education, nursing care, and other areas, has been taking various steps to secure new recruits, including having new employees produce a pamphlet detailing its operations. The brochure reports on the company for future employees and articles on such topics as female employees who are raising children. A Benesse official said the pamphlet gives potential employees a chance to gain a deeper understanding of the company and experience a process akin to actual work. He said the participants maintain good teamwork. 3. Software developer Cybozu formed the Nintendo DS Club for playing video games 2 years ago on the condition that members come from all three divisions in order to achieve employee interaction transcending divisional walls and keep job turnover low. Members are free to decide the nature of activities and can get monetary aid from the company to fund them.		

		ideas, the public sector should place greater emphasis on providing better vocational training. But there is a possibility that during the process in which the information is collected by local governments, such young people may feel that their privacy has been invaded and that they are being stigmatized by society. The government should carefully consider how it can help those in need without stirring negative feelings.		
8	*Subsidies to make "freeters" full-timers* (countermeasures to freeters/subsidies)	The Japanese government plans to give companies a subsidy of up to ¥1 million for each nonstudent part-timer, or "freeter," they hire as a full-time employee. The subsidies will be granted to companies hiring part-timers aged between 25 and 34 as regular employees. The planned subsidies are part of Prime Minister Taro Aso's plan to boost the employment of young people, including part-timers, who are expected to face difficulty finding stable jobs in the wake of the global financial turmoil, the sources said. The government will grant about ¥500,000 to large firms and about ¥1 million to midsize and small companies per hire, they said, noting the program will have a 3-year limit.	9	*Kyodo News*, Oct. 22, 2008
9	*New plan aims to reduce "freeters" by 110,000* (countermeasures to freeters/training)	A job promotion program has been developed to reduce the number of job-hopping "freeters" by 110,000 by fiscal 2010 to spur the economy. The program features steps to help the job hoppers acquire regular full-time jobs and will offer training so they can improve their professional skills, government officials said. The employment plan emphasizes the acceleration of the new "job card system" that began earlier this month. The system distributes job cards to freeters and single mothers to help them find regular full-time jobs in a full-fledged manner. The government also plans to up the number of cardholders to 500,000 over the next 3 years.	9	*Kyodo News*, April 23, 2008

Table 7.2. (continued)

No.	*Article title (points of contents)*	*Summary of the articles*	*Related statistics no. in Table 7.1.*	*Source*
6	*40% of new adults fret harsh lives* (younger workers are not focused on profession)	More than 40% of those entering adulthood believe their lives will be worse than their parents' lives, a survey by a bridal information company said. According to the O-net Inc. survey conducted on 800 people born between April 1987 and April 1988, 43% of respondents believe they will have to lead tougher lives than their parents did and 44% believe their children's lives will be even worse than theirs. Asked how they want to spend their lives in the future, 18% said they wanted to seek financial affluence and 51% said they prefer lives that allow them to get along with friends and family members, the survey said. While the ratio of those who fear becoming "freeters" (job-hopping part-timers) dropped by half to 26% compared with 2 years ago, only 18% said they would persevere in finding the exact job they want. That indicates most young people are not strictly focused on finding a profession, the survey said.	11, 12	*Kyodo News*, Jan. 13, 2008
7	*Nurturing young people* (counter measures to freeter problem/job cards and trial employment system)	The Japanese government has adopted new general principles for nurturing young people, especially those who shut themselves in at home and those not engaged in education, employment, or training (NEETs). The document proposes a wider use of "job cards" containing employment history and assessments by former employers to help young people find new jobs, and the introduction of a "trial employment system," under which public employment security offices (PESOs) would help young people get full-time jobs after a trial period. While these are good	9	*Japan Times*, Jan. 7, 2009

Table 7.2. (continued)

No.	*Article title (points of contents)*	*Summary of the articles*	*Related statistics no. in Table 7.1.*	*Source*
		Lending will be increased to people who attend job-training courses. The plan also will revive employment-introduction agencies, tentatively called Home Town Hello Work, together with prefectural and municipal governments in the districts where they have been shut down.		
10	*Fallout of population decline: Rural universities feel pinch of lower enrollments* (countermeasures to freeters/reeducation plan for freeters)	Japan's 18-year-old population, which peaked at 2.05 million in fiscal 1992, dropped to only 1.37 million in fiscal 2005 and is projected to fall to 1.24 million in fiscal 2008. Meanwhile, a policy of deregulation since the 1990s has allowed more colleges and departments to start up, with the number of private universities rising from 379 in fiscal 1992 to 559 this year, a 147% increase. Some say that universities should look at potential new demand at home, adding that there are huge untapped educational needs among people like the so-called freeters and others hoping to upgrade their skills and move into new careers, as well as women who want to reenter the labor market after raising children. While Japanese universities have traditionally defined their markets as people in the 18–22 age brackets, they can start to consider corporations as new customers, because the companies need customized training programs to help their employees upgrade skills.	9	*Japan Times*, Aug. 17, 2007

Recruiting Schema Problems

Increasing nonregular workers is supposed to be the company's best option in accomplishing cost reduction. This option is, however, the absolute end of the "Japanese dream" of lifetime employment with seniority. In addition, the freeter problem is much harder for older people and the unemployed who are caught up in the spiral of poverty. These negative trends are perhaps a by-product of the mentality that workers are merely economic production units. This mechanical way of treating workers who cherish deductive skills should be changed.

Companies have recently been turning to new tactics to nurture fresh recruits. In order to alleviate the high turnover of new employees, companies have been exerting communication training, which has been neither successful nor effective. One reason for the failure might be the task process in actual workplaces, which are not collaborative or interactive between and among workers. Thus, the workplace and its environment should be improved toward collaborative or ecologically dependent workplaces.

The governmental (corporate as well) countermeasures (subsidies, personal data-enlisting system, etc.) are analytic or static strategies and insufficient. Moreover, those countermeasures sometimes create severe side effects such as social costs and extra budgets, heavy tax burdens, privacy problems, and others. The countermeasures, therefore, should be synergetic or synthetic strategies.

Thus, the problems are categorized as follows: (a) deductive way of reasoning, (b) noncommunicative working environment, (c) fixed task that requires specific skill, (d) behaviorism or objectivity, and (e) static strategy.

Recruiting Schema Requirements

When we stand on the view as stated earlier that the efficiency of business is based upon the accumulation of individual, synergetic teamwork effort, the organizational synergy requires team coordination, and communication becomes an important factor in designing work systems. According to sources that are supported by government-issued statistics, the following are required by both employer and employee to alleviate the earlier issues:

1. Rationalization and abstraction ability realized by worker-oriented environment (Institute for Science of Labour, 2008b)
2. Communication in collaborative task process (Institute for Science of Labour, 2008a; Cabinet Office of the Japanese Government, 2005)
3. Self-realization in discretionary task performance (Institute for Science of Labour, 2008a)
4. Incentives realized by intentionality in performing tasks (Cabinet Office of the Japanese Government, 2008)
5. Flexible job design and performance through dynamic analysis (Cabinet Office of the Japanese Government, 2008)

It is also noted that the governmental countermeasures are not sufficient and sometimes create severe side effects such as privacy problems. Workplace designers should aim at sufficient countermeasures.

ISO 6385-2004 FOCAL POINTS

In recent years business global operations have become quite common, so more and more, workers are joining work teams whose members are from various nations. For an effective design of work systems, we need a basic framework for those who deal with the issues of work systems and working situations that facilitate mutual understanding among people with different cultural backgrounds. Since ISO is a worldwide federation of national standards, the standards can be one of the bases for mutual norms over, at least, member nations. As stated in ISO 6385-2004, described as "ergonomic principles in the design of work systems," this international standard can provide a basic framework for professionals and other people who deal with the issues of work systems and working situations.

ISO 6385-2004 Overview

Work system modeling must meet special constraints in order for the jobs to be effectively performed in workplaces. Current macroergonomics-based work analysis and design methodologies are well defined (Hendrick & Kleiner, 2001) and prepared in terms of three crucial workplace subsystems: ambient environment, workers and equipments, and operations. These terminologies and methodologies are realized with such international standards as ISO 6385-2004 and other related or referenced standards.

The standard provides the basic guidelines for the design of work systems with the definition of relevant basic terms. For example, it defines the term "work system" as a work system that involves a combination of people and equipment within a given space and environment and the interactions between these components within a work organization. It is recommended to designers for good communication that this definition and other terms be used to mean what a technical term connotes.

The second section of the standard defines 17 terms, which include the allocation of system functions, design, ergonomics study of human factors (systematic study concerned with the understanding of the interactions between human and other elements of a system), job (organization and sequence in time and space of an individual's work tasks by one worker within a work system), system function, work environment (physical, chemical, biological, organizational, social, and cultural factors surrounding a worker), work equipment, worker operator, work fatigue, work organization (sequence and interaction of work systems fitted together to produce a specific result), work process, workstation, work strain, work stress due to external load, workspace, work system (system comprising one or more workers and work equipment acting together to perform the system function in the work space and in the work environment under the conditions imposed by the work tasks), and work task.

The third section describes the design principles as "designing work systems," which consists of seven clauses: general principles; the work system design process; formulation of goals (requirements analysis); analysis and allocation of functions; design concept; detailed design; and realization, implementation, and validation. It is recommended for practitioners and those who are concerned with effective HRM that these contents be reviewed and implemented in an individual's work systems in an organization. Some of them will be introduced here. For details, refer to ISO 6385-2004.

Most of the HRM issues previously stated include workplace difficulties, such as less loyalty and low motivation of workers, the management's frustration, and the shift to the use of part-timers who are not well qualified. These are derived from the relatively common idea that "workers are part of mechanisms within a human–machine system" (see Figure 7.1; Saito, 1998), which was originally an excerpt from the first edition of ISO 6385.

Figure 7.1 depicts a worker as a mechanistic part in a human–machine system in the human–machine framework. Saito stresses the importance of the shift from mechanical to environmental perspective in terms of workers in a human working environment system (see Figure 7.2, which is an excerpt from the second edition of ISO 6385-2004).

ISO Focal Points

ISO 6385-2004 also asserts important points to be considered for work system design as follows: optimal working conditions, workers' involvement, and work system evaluation parameters. These points are useful for practitioners to follow when they design work systems.

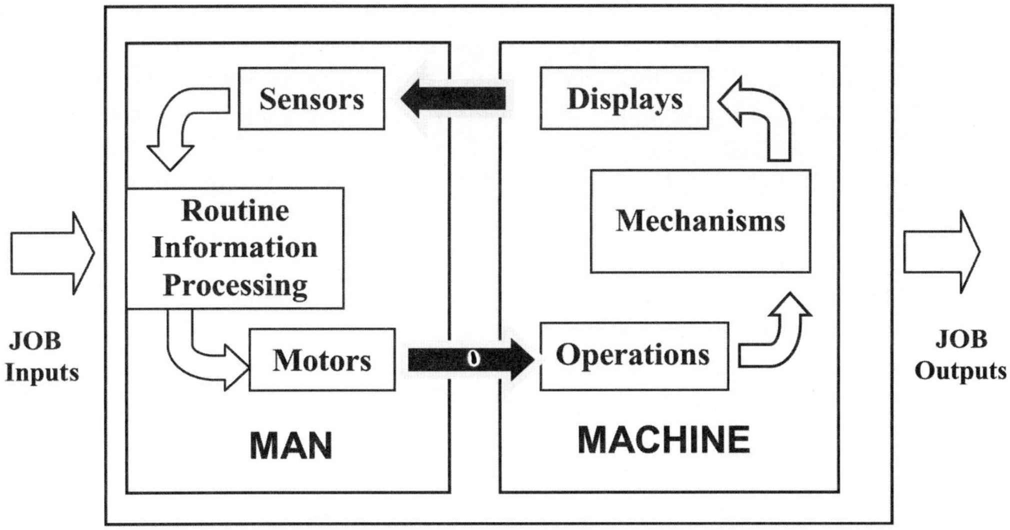

Figure 7.1. Worker as a mechanistic cog in the human–machine framework.

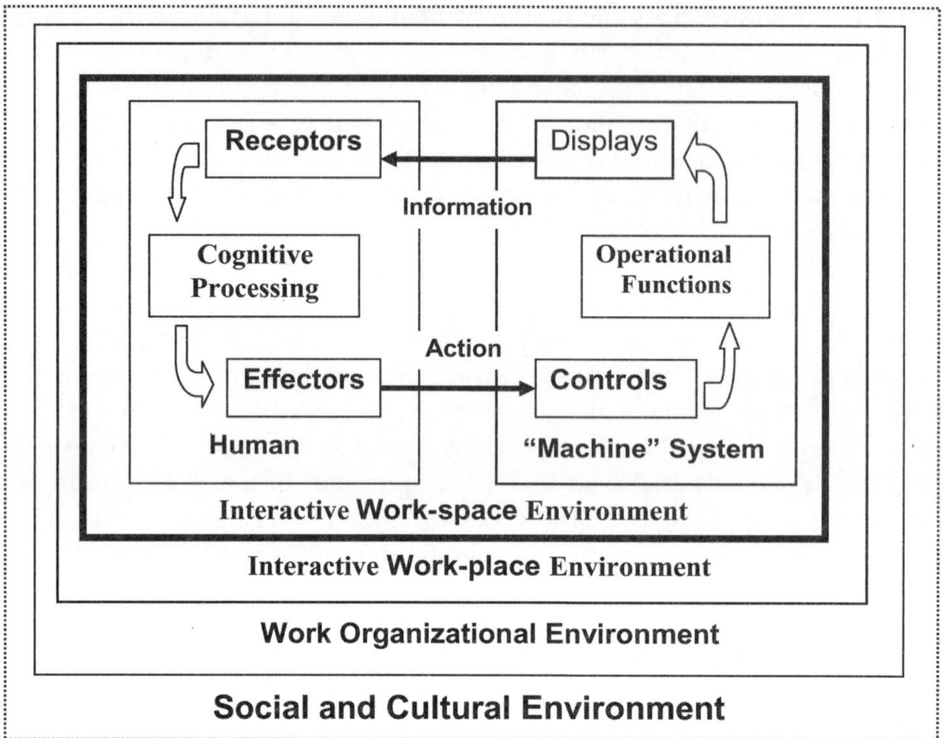

Figure 7.2. Worker as a system actor in the human–environment framework.

1. *Optimal working conditions.* Imbalance between human ability and job demand should be eliminated by adjustments between "fitting the job to the worker" and "fitting the worker to the job" parameters, which are indicated in Figure 7.3. Note that "fitting the job to the worker" has been recommended by ergonomists for a long time, and its reverse strategy is brought here by the means of training and selection to compromise the difficulty of design.
2. *Workers' involvement.* Workers' involvement in the system development from the design stage includes the following:

    a. Task design: promotion of task performance
    b. Design of workspace: postural stability and postural mobility
    c. Interface design: physical or mechanical and cognitive aspects
    d. Design of work environment: physical, chemical, and biological conditions
    e. Job design: work process design, job enlargement, job enrichment, and voluntary job rotation
    f. Social and cultural environments

These designs have been and are done by management, engineers, or specialists other than workers who are supposed to work. Note that the essential term is workers' involvement in practice. This study strongly recommends to exercise a participative method.

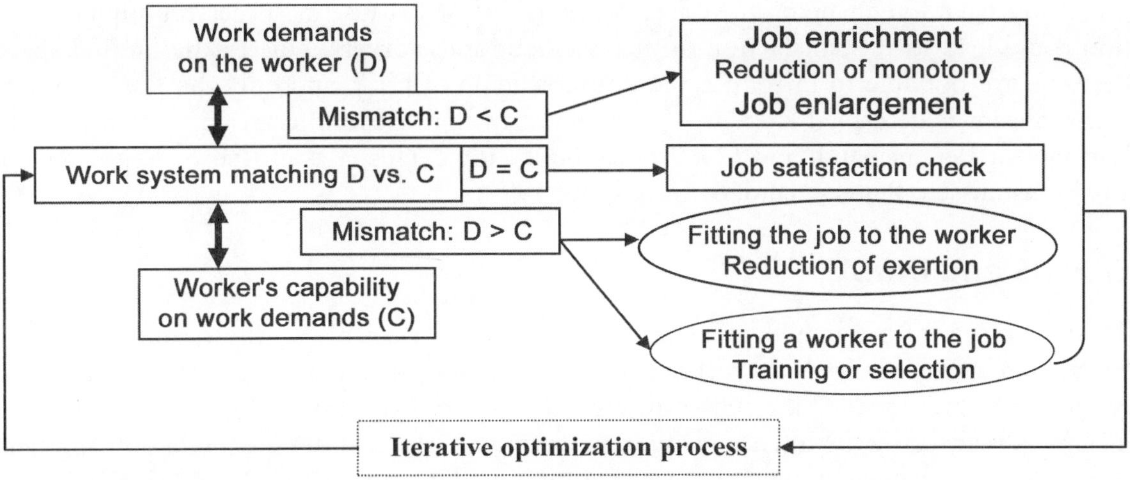

Figure 7.3. Optimization process in the work system matching framework.

3. *Work system evaluation parameters.* The work system evaluation parameters include the following:

   a. Health and well-being—physiological, psychological, and subjective
   b. Safety reliability, errors, unsafe behavior, near-miss
   c. Performance—quality, quantity

ECOLOGICAL APPROACHES TOWARD INTEGRATED HUMAN RESOURCE MANAGEMENT

Aiming at the solutions to the issues, the thesis of this section introduces ecological approaches that are relevant methods to enhance the effectiveness of HRM. They are theoretically no longer new, and there are many publications found in the literature since the 1950s. But the implementation of these concepts and philosophy is scant, and they still have not been implemented in workplaces that need it to be done.

However, ISO standards or macroergonomics technology does not typically result in models that meet cognitive constraints, largely because the environmental influences on cognition are investigated in a piecemeal fashion (Kirlik, 1995). Such a conception only serves to license the underlying gap between psychological modeling and work system demands, which abstracts away much of the complexity of the environment and design practice. To overcome this problem, ecological task analysis (ETA) is introduced for suggesting how interfaces and interactions can be manipulated to alleviate certain cognitive demands (Rasmussen & Pejtersenand, 1995). We strongly advocate that their ETA concept is quite a useful approach to enhancing the effectiveness of HRM because its successful implementation could lead to the satisfaction of management and workers. Due to the limited space here, we are not able to introduce the entire scheme of ETA, and only the framework of their concept is described later. It is recommended that those who are interested refer to Rasmussen, Pejtersenand, and Goodstein's books for details (Rasmussen & Pejtersenand, 1995; Rasmussen, Pejtersenand, & Goodstein, 1994).

EAHMS and IHRM

To answer part of the employment problems discussed earlier (i.e., the rationalization or abstraction ability realized by worker-oriented environment, improving communication in collaborative task process, the self-realization in discretionary task performance, improving incentives realized by the intentionality in performing tasks, and flexible job design and performance through dynamic analysis), we have attempted to utilize EAHMS and IHRM theory and practice for a solution that could satisfy the employers and employees. As a practical methodology, EAHMS or IHRM is characterized by those methods as job abstraction, closed-loop feedback, bottom-up decision making, and a knowledge database system (Ikeda, 2003).

As a result, the present study formed the points that characterize EAHMS and IHRM methods as follows:

1. Jobs abstraction for intersectional work analysis and integrated management system for worker-oriented environment
2. Closed-loop feedback system between and among workers and managers for collaborative task process
3. The bottom-up decision-making system for well-motivated workers (employees) for flexible job design and performance through dynamic analysis
4. Information and communication technology-based knowledge sharing system, especially knowledge database systems for self-realization in discretionary task performance

"Improving and upgrading incentives and intentionality" is realized by intentionality in performing tasks through all these processes of points 1 through 4, accordingly.

These points with others should improve the well-being of managers (employers) and workers (employees). Table 7.3 represents these ideas in an integrated fashion from the relationship view of human resource recruitment and human resource management.

Focal Points of Ecological Approach With EAHMS and IHRM

This section describes ecological interface design (EID), which is the background theory of EAHMS and IHRM, and the focus points of ecological approach will be summarized with the discussions. EID has been elaborated by Rasmussen and his associate, who asserted a three-layered human cognitive behavior model, in other words, skill-based, rule-based, and knowledge-based, known as SRK, with a five-layered workplace abstraction hierarchy, also

Table 7.3. Human Resource Recruitment and Management Relationship

	Human resource recruitment		Human resource management
	Target	*Requirement*	*EAHMS/IHRM*
1	Rationalization ability	*Worker-oriented environment*	Jobs abstraction for intersectional work analysis and integrated management system
2	Communication ability	*Collaborative task process*	Closed-loop feedback system between and among workers and managers
3	Self-realization	*Discretionary task performance*	Bottom-up decision-making system for well motivated workers (employees)
4	Incentives	*Intentionality in performing tasks*	Intentionality in performing tasks through all these processes 1, 2, 3, 5
5	Flexible job design and performance	*Dynamic analysis*	Information and communication technology-based knowledge sharing system

known as means-ends relations, that includes (a) purpose and values (constraints posed by environment); (b) priority measures (flow of mass, energy, information, people, and monetary values); (c) general work activities and functions; (d) specific work process and physical processes equipment; and (e) appearance, location, and configuration of material objects in top-down reasoning order, respectively (for details, refer Flach, 1995). Upper layers are the purpose of the lower layers. Lower layers are, in contrast, means of accomplishing the upper layers' ends.

Thus their idea can be characterized by a two-axis layer system of SRK and means-ends hierarchy. Each layer accompanies proper language, such as natural, logic, systems modeling, computer programming, physics, and so on. All these are integrated into an ecological interface design (EID) diagram. Rasmussen and Pejtersenand (1995) clarify that the ecological approach theory focuses mainly on the points of work systems design following the five orientations as follows:

Workers Oriented

For many modern work systems, stable or fixed work procedures are not the norm and many tasks are discretionary. Thus the object of modeling can no longer be the "task" but must include all the features of the work environment and the interpretation of these features by the workers. The interaction of task and the environment also creates the task ad hoc. In this sense, ecological approach should be worker oriented.

Collaborations Oriented

Institutional constraints (intentions), propagating top-down, will interact with criteria that are dependent on the locality, situation, and person. Therefore, in many cases, actors judge the intentions of cooperators through formal and informal communication channels (e.g., body language and other nonverbal cues).

Decision making is more manageable when decomposed into subroutines for linking different processes for communicating with cooperators. This is important because a complex decision task may be shared not only by a number of cooperating team members but also across time with procedure designers, system programmers, and computers. In this sense, the ecological approach should be collaboration oriented.

Discretionary Work Oriented

Traditionally, the identification of work requirements in a particular situation is done in terms of a task analysis to produce a description in terms of sequences of actions. This is no longer an adequate approach when tasks are discretionary. Instead, activities must be decomposed and analyzed in terms of a set of problems to be solved or a set of prototypical work situations that represent recurrent, natural islands of activity that have reasonably well-defined boundaries. In this sense, the ecological approach should be discretionary work oriented.

Intentionality Oriented

In a loosely coupled workplace such as an office, a traditional manufacturing workshop, or a hospital, the functional coordination depends on the intentionality brought to work by the actors of the system. In Rasmussen and Pejtersenand (1995), a city has been taken as an example of a loosely coupled work system (power plant as a tightly coupled system) and been used to illustrate the role of the functional and intentional constraints represented by the means-ends hierarchy.

In general, a loosely coupled workplace is more common today. In this sense, the ecological approach should be intentionality oriented.

Dynamics Oriented

The context and the perspective taken by the worker will change dynamically during a decision task in terms of strategy. In this sense, ecological approach should be dynamics oriented. This is well illustrated in Rasmussen and Pejtersenand (1995) as presented in the following:

> *A mental strategy* is an abstraction describing one consistent reasoning approach characterized by a certain kind of mental representation and a particular interpretation of observations. The concept of a *formal* strategy serves to formalize the information processes, their information requirements, and the cognitive load involved. The *natural* strategies applied in actual work situations, for several different reasons, involve very frequent shifts among such formal strategies caused by very subjective and situation-dependent factors. One reason to shift between strategies is their very different resource requirements with respect to time taken, type of mental model available, number of observations necessary, and so on. Shift in strategy, i.e., dynamics, is a very effective way to circumvent local difficulties along the path of work.

Another reason for a shift of strategy (i.e., dynamics) in a particular work scenario is that the diagnostic objective will change during a session. Initially a medical doctor or a process operator will be concerned with the question of whether he or she is confronted with a need for rapid compensatory action, that is, the user is concerned with the potential consequence of the present state. Next, the operator will be concerned with the choice of a function to control for compensation. Then he or she will be concerned with the correction of the present abnormality and, finally, may be concerned with prevention of a repetition. This means that a diagnostician may need to shift between strategies and thus change perspective in the interpretation of constraints several times during a task. An important implication of this discussion is that an information system should support all the strategies relevant for a task to allow users to shed mental workloads by shifts among strategies.

It should be noted that the human resource management requirements, EAHMS and IHRM methods, and the ecological approach are closely related in terms of the five orientations mentioned earlier.

DISCUSSION

From the personnel management statistics as stated earlier, corporate management tended to increase the number of nonregular employees compared with regular ones during the years from 2002 to 2007. A skill of rationalization is the most required skill for employees, compared with general line or assembly line job skills. On the other hand, workers prefer more mental to physical richness in their jobs. They also prefer free time to higher salary, as the trend appeared during the years from 1997 to 2007. In short, these statistics indicate that both management and workers want to have somehow a humanized and high-quality working environment, but each group views it from a different standpoint. Management seeks efficiency through workers, and workers seek more freedom and job enrichments. It is necessary for HRM to compromise so that the desires for enhancing the effectiveness of the HRM system on both sides are met.

The ISO ergonomics standard requires optimal working conditions, workers' involvement in system development, and others. The workers' involvement in workplace design is very important for achieving the needs of contemporary society due to, at least, two reasons: mutual understanding between workers and management, and enhancing workers' motivation through their participation in designing their own jobs.

The identified requirements for the current work system analysis and design could be abstraction hierarchy with shift of languages and a closed bottom-up loop work system. To meet with these requirements we need a methodology to implement them in workplaces. The ecological approaches to work systems design could be a feasible method to achieve the enhancement of HRM systems.

At the turn of the 20th century, we have globally experienced dramatic changes in technological and economic factors, which affected organizational behaviors on their management models. The changes in HRM models certainly affected the work behavior and well-being of people as part of a work system. To discuss the solution of recruiting problems in the 21st century, we should start with 20th century human resource management and its problems, among which human resource recruiting schema is, we would think, the root of HRM problems.

Next we could carry on to the ecological approach to human–machine systems with its application to human resource management, human resource development, and finally, phenomenology as the foundation of all the theories and methods.

Ecological Approach and Affordance

The definition of ecological approach is mapping "affordance" to the personal perception so that the perception and environment are well interrelated and interact with each other effectively. In this sense, the main idea of ecological approach is subjective (or intersubjective in phenomenological and reductive sense), contrary to the behaviorism that utilizes a deductive and an objective way of reasoning. Affordance is the functional property of environment (Gibson, 1979). There, direct perception is the specificity of information in sensory arrays that should be mapped to affordance. The ecological approach and affordance are characterized as follows:

1. Human–environment interdependency (interactive abstraction)
2. Intersubjective (common) perception or invariants
3. Subjectivity
4. Nonbehaviorism way of analysis and intentionality
5. Gestalt principle in synthetic way of analysis

These characteristics were originally asserted by J. J. Gibson in his works, which are inspired by the phenomenological approach advocated by Husserl (1982).

Twentieth-Century Human Resource Management and Its Problems

We should revisit the 20th-century recruiting schema, which targets following factors:

1. Deductive way of reasoning
2. Environment-independent skills (in noncommunicative working environment)
3. Fixed task with specific skill
4. Behaviorism and objectivity
5. Static or antisynthetic (analytic) approach

The recruiting schema problems are the "root of all evil" in terms of working environment obstacles. For example, the obstacles in terms of work design (Campion, Mumford, Morgeson, & Nahrgang, 2005) might be caused by the problems stated earlier (see Table 7.4).

Table 7.4. Obstacles Caused by Twentieth-Century Recruiting Schema

	Obstacles	*Causes*
1.	Work design influences multiple outcomes	Static or antisynthetic (analytic) approach
2.	There are trade-offs between different work design approaches	Static or antisynthetic (analytic) approach
3.	It is difficult to choose appropriate unit (jobs, tasks, duties, etc.) of analysis	Deductive way of reasoning Environment-independent skills Fixed task with specific skill Static or antisynthetic (analytic) approach
4.	It is difficult to predict the nature of a job before it exists	Environment-independent skills Fixed task with specific skill Behaviorism/objectivity Static or antisynthetic (analytic) approach
5.	Individual differences complicate job design	Environment-independent skills Fixed task with specific skill Behaviorism/objectivity Static or antisynthetic (analytic) approach
6.	Job enlargement can occur without job enrichment	Environment-independent skills Static or antisynthetic (analytic) approach
7.	New job needs to be created as part of growth or downsizing	Deductive way of reasoning Fixed task with specific skill Behaviorism/objectivity Static or antisynthetic (analytic) approach
8.	Long-term effects may differ from short-term effects	Environment-independent skills Static or antisynthetic (analytic) approach

Twenty-First-Century Human Resource Recruiting Schema

In contrast to the 20th-century recruiting schema, 21st century recruiting schema identifies human resource requirements as follows:

1. Abstraction and rationalization skills
2. Communication skills
3. Self-realization
4. Intentionality
5. Dynamic or synthetic design skills

To sum up these points, the "versatility" of a worker is the best word and can be measured mainly by abstraction or rationalization and synthetic skill. For example, the human resource department staff often utilizes the Fermi test (a test of synthetic and abstractive ability) for interviews in order to verify the competency of the interviewee in terms of versatility and skills that were mentioned earlier.

Here is a typical Fermi test example, an excerpt from http://mathforum.org/workshops/sum96/interdisc/classicfermi.htm.

Question: How many piano tuners are in New York City?

Answering process:
Surely the number of piano tuners in some way depends on the number of pianos. The number of pianos must connect in some way to the number of people in the area.

Approximately how many people are in New York City?
10 million.

Does every individual own a piano?
No.

Would it be reasonable to assert that "individuals don't tend to own pianos; families do"?
Yes.

About how many families are there in a city of 10 million people?
Perhaps there are 2 million families in NYC.

Does every family own a piano?
No.
Perhaps one out of every five does.
That would mean there are about 400,000 pianos in NYC.

How many piano tuners are needed for 400,000 pianos?
Some people never get around to tuning their piano; some people tune their piano every month. If we assume that "on the average" every piano gets tuned once a year, then there are 400,000 "piano tunings" every year.

How many piano tunings can one piano tuner do?
Let's assume that the average piano tuner can tune four pianos a day. Also assume that there are 200 working days per year. That means that every tuner can tune about 800 pianos per year.

How many piano tuners are needed in NYC?
The number of tuners is approximately 400,000/800 or 500 piano tuners.

Ecological Approach to Human–Machine (Environment) Systems and Integrated Human Resource Management

From the integrated human resource management point of view, the following orientations are indispensable:

1. Worker oriented
2. Collaboration oriented
3. Discretionary oriented
4. Intentionality oriented
5. Dynamics oriented

These factors are asserted by Kirlik, Rasmussen, and Flach in their influential papers (Flach, 1995; Kirlik, 1995; Rasmussen & Pejtersenand, 1995), which were inspired by ecological psychology theory (Gibson, 1979) based on phenomenology (Husserl, 1982).

Ecological Approach to Human Resource Development (Organizational Learning)

From the human resource development point of view, organizational learning theory targets the following methods:

1. Challenging mental model
2. Shared vision
3. Personal mastery
4. Team learning
5. Systems thinking

First, Argyris and Schön (1996) distinguish between single-loop and double-loop learning, which are related to Gregory Bateson's concepts of first- and second-order learning, which in turn are closely related to the phenomenological approach (Husserl, 1982). In single-loop learning, individuals, groups, or organizations modify their actions according to the difference between expected and obtained outcomes. In double-loop learning, the entities (individuals, groups, or organization) question the values, assumptions, and policies that led to the actions in the first place; if they are able to view and modify those, then second-order or double-loop learning has taken place. Double-loop learning is the learning about single-loop learning. This idea has been refined and extended as the learning of organization theory asserted by Senge (1990). Learning organization methodology consists of the following five disciplines.

Challenging Mental Models

Mental models are the terms given to ingrained assumptions held by individuals and organizations. To become a learning organization, these mental models must have been challenged. Individuals tend to espouse theories, which they intend to follow, and theories-in-use, which is what they actually do.

Shared Vision

The development of a shared vision is important in encouraging the workforce to learn as it creates a common identity that can provide focus and energy for learning. The most successful visions build on the individual visions of the employees at all levels of the organization. The creation of a shared vision is likely to be hindered by traditional structures where a company vision is imposed from a higher authority. As a result, learning organizations tend to have flat, decentralized organizational structures.

Personal Mastery

Personal mastery is the commitment by an individual to the process of learning. There is a competitive advantage for an organization whose workforce can learn quicker than the workforce of other organizations. Individual learning is acquired through staff training and development. However, learning cannot be forced upon an individual if he or she is not receptive to learning.

Team Learning

Learning organizations have structures that facilitate team learning with features such as boundary crossing and openness. Team learning requires individuals to engage in dialogue and discussion, therefore it is important that team members develop open communication, shared meaning, and understanding. Learning organizations also have excellent knowledge management structures, which allow the creation, acquisition, dissemination, and implementation of this knowledge throughout the organization.

Systems Thinking

The idea of the learning organization originally developed from a system thinking theory based on system dynamics theory. This is a conceptual framework that allows people to study businesses as bounded objects. Learning organizations employ this method of thinking when assessing their company and develop information systems that measure the performance of the organization as a whole and of its various components.

Phenomenology

Finally, we must revisit the fundamental theory of all the methods stated earlier, that is, "phenomenology" advocated by E. Husserl, who was inspired by Gestalt psychology in the early 20th century. The basic ideas of phenomenology are the following:

1. Reduction or reflective analysis
2. Intersubjectivity
3. Subjectivity
4. Intentionality and nonbehaviorism
5. Gestalt view of analysis

Thus, the triad of recruiting schema (human resource recruitment), ecological approach (human resource management), and organizational learning (human resource development) is successfully and coherently completed (see Table 7.5).

CONCLUSION

First, this chapter introduced the HRM problems and personnel management issues from the results of analyzed personnel management statistics reported by governmental agencies in Japan. These results have suggested that workers are nowadays treated according to their status in the company and the lower-end workers rarely express their opinions, causing them to be stressed. Employees, therefore, become less motivated and less loyal to their team or company. Management also becomes frustrated and relies heavily on part-timers who are not well qualified, at least in Japan. These problems and issues are often affected by working environment factors.

To locate the root cause of HRM problems, it is referenced and analyzed from the working environment point of view using news articles that are supported by official and public statistics such as labor white papers (or white books) and other private statistics issued both in Japan and other Western countries. These have suggested that a corporate recruiting schema favors temporary employment, especially for younger workers. It is also noted that senior workers have not been offered relevant jobs to their skills and experience. The tendency toward just temporary employment could actually increase the freedom of choice for younger employees. However, unfair age discrimination adds friction to the relationship between the employer and employee. The essential solution to these problems might be in the enhancement of human resource capacity and capability. These abilities are summarized as recruiting schema requirements, which include rationalization ability, communication ability, self-realization, incentives realized by intentionality, and flexible job design. It is also noted that the governmental countermeasures are not sufficient and sometimes create severe side effects such as privacy problems.

Table 7.5. IHRM (HRR–HRM–HRD) Triad Relationship

	Human resource recruitment		*Human resource management*		*Human resource development*	*Fundamental theory*	
	Target	*Requirement*	*Ecological approach*	*EAHMS/IHRM*	*Organizational learning*	*Affordance*	*Phenomenology*
1	Rationalization ability	*Worker-oriented environment*	*Worker oriented*	Jobs abstraction for inter- sectional work analysis and integrated management system	Challenging mental model	Human environment interactive or abstraction	Reduction with reflective analysis
2	Communication ability	*Collaborative task process*	*Collaborations oriented*	Closed-loop feedback system between and among workers and managers	Shared vision	Intersubjectivity (invariants)	Intersubjectivity
3	Self-realization	*Discretionary task performance*	*Discretionary work oriented*	Bottom-up decision-making system for well-motivated workers (employees)	Personal mastery	Subjectivity	Subjectivity
4	Incentives and intentionality	*Intentionality in performing tasks*	*Intentionality oriented*	Incentives realized by intentionality in performing tasks through all these processes 1, 2, 3, 5	Team learning	Intention with non-behaviorism	Intention with nonbehaviorism
5	Flexible (dynamic/synthetic) job design and performance	*Dynamic analysis*	*Dynamics oriented*	Information and communication technology-based knowledge sharing system	Knowledge database	Gestalt synthesis	Gestalt synthesis

Second, global business operations call for a framework that can internationally provide mutual understandings to HRM system designers when they work. ISO is a worldwide federation of national standards. The standards can be one of the bases for mutual norms. It is recommended to use ISO 6385-2004, "ergonomic principles in the design of work systems," as a basic framework. It asserts important points to be considered for work system design as optimal working conditions, workers' involvement, and work system evaluation parameters. These points are useful for practitioners to follow when they design human work systems and their organization.

Third, ecological approaches were examined to solve part of the HRM issues and recommended to apply to the designing of tasks, jobs, human–machine systems, and HRM organization. As a result, the present study formed the points that characterize the EAHMS and IHRM methods as follows: (a) a bottom-up decision-making system for well-motivated workers (employees) for realizing rationalization ability, (b) a closed-loop feedback system between and among workers and managers for communication, (c) a jobs abstraction for intersectional work analysis and an integrated management system for flexible job design, and (d) an information and communication technology-based knowledge sharing system, especially a knowledge database system for self-realization. The four factors are combined to realize intentionality. These points with others improve the well-being of managers (employers) and workers (employees).

Finally, we have discussed the solution of human resource recruiting problems in the 21st century, starting with the 20th-century recruiting schema and its problems with our definition of ecological approach and affordance, which leads to 21st-century human resource recruiting schema. It identifies human resource requirements as follows: abstraction and rationalization skills, communication skills, self-realization, intentionality, and dynamic or synthetic design skills. From an integrated human resource management point of view, the following orientations are indispensable: worker oriented, collaboration oriented, discretionary oriented, intentionality oriented, and dynamics oriented. From a human resource development point of view, organizational learning theory targets the following methods: challenging mental model, shared vision, personal mastery, team learning, and systems thinking. It is proposed that the triad of recruiting schema (human resource recruitment), ecological approach (human resource management), and organizational learning (human resource development) be used for recruiting. However, further implementation of EAHMS or IHRM by HRM practitioners may require detailed tools whose development is under preparatory study and to be presented elsewhere.

REFERENCES

Argyris, C., & Schön, D. (1996). *Organizational learning*. Reading, MA: Addison-Wesley.

Cabinet Office of the Japanese Government. (2005). *Consciousness survey concerning the youth's social independence*. Tokyo: Japanese Cabinet Office.

Cabinet Office of the Japanese Government. (2008). *Public opinion poll concerning national life*. Tokyo: Japanese Cabinet Office.

Campion, M. A., Mumford, T. V., Morgeson, F. P., & Nahrgang, J. D. (2005). Work redesign: Eight obstacles and opportunities. *Human Resource Management*, *44*(4), 367–390.

Flach, J. M. (1995). The ecology of human-machine systems: A personal history. In J. M. Flach, P. Hancock, J. Caird, & K. Vincente (Eds.), *Global perspectives on the ecology of human-machine systems* (Vol. 1, pp. 121–156). Hillsdale, MI: Lawrence Erlbaum Associates.

Gibson, J. J. (1979). *The ecological approach to visual perception*. Boston, MA: Houghton Mifflin.

Hendrick, H. W., & Kleiner, B. M. (2001). *Macroergonomics—An introduction to work system design*. Santa Monica, CA: Human Factors and Ergonomics Society.

Husserl, E. (1982). *Ideas pertaining to a pure phenomenology and to a phenomenological philosophy—first book: General introduction to a pure phenomenology*. The Hague, Netherlands: Nijhoff.

Ikeda, Y. T. (2003). The importance of participative and ecological approaches to effective and efficient work system designing. In *Proceedings of the 10th Annual Meeting and Conference of Ergonomic Society of Taiwan* (pp. 43–50). Taiwan: Ergonomics Society of Taiwan.

Ikeda, Y. T., & Masuzawa, Y. (2009). Ergonomic related ISO and its effective implementation utilizing EAHMS analysis for integrated HRM—an empirical study in workplaces. In *Proceedings of the International Conference on Enhancing Resource Effectiveness in the Workplace* (pp. 93–101). Kathmandu, Nepal: IJIE.

Institute for Science of Labour. (2006). *Survey on personnel strategy and the worker's consideration under diversified employment*. Tokyo: Institute for Science of Labour.

Institute for Science of Labour. (2008b). *Survey on workers' skill development and securing in manufacturing industries*. Tokyo: Institute for Science of Labour.

Institute for Science of Labour. (2008a). *The employee's consideration and problem of manpower management*. Tokyo: Institute for Science of Labour.

Japanese Ministry of Health, Labour, and Welfare. (2008). *Labour economics white paper*. Tokyo: Japanese Ministry of Health, Labour and Welfare.

Japanese Statistics Bureau. (2007). *Labour force survey*. Tokyo: Japanese Ministry of Internal Affairs and Communications.

Kirlik, A. (1995). Requirements for psychological models to support design: Toward ecological task analysis. In J. M. Flach, P. Hancock, J. Caird, & K. Vincente (Eds.), *Global perspectives on the ecology of human-machine systems* (Vol. 1, pp. 68–120). Hillsdale, MI: Lawrence Erlbaum Associates.

Masuzawa, Y., & Ikeda, Y. T. (2009). EAHMS/IHRM approach for solving problems in current corporate recruiting schema. In *Proceedings of the International Conference on Enhancing Resource Effectiveness in the Workplace* (pp. 82–92). Kathmandu, Nepal: IJIE.

Rasmussen, J., & Pejtersenand, A. M. (1995). Virtual ecological task analysis. In J. Flach, P. Hancock, J. Caird, & K. Vincente (Ed.), *Global perspectives on the ecology of human-machine systems* (Vol. 1, pp. 121–156). Hillsdale, MI: Lawrence Erlbaum Associates.

Rasmussen, J., Pejtersenand, A. M., & Goodstein, L. P. (1994). *Cognitive systems engineering*. New York, NY: John Wiley & Sons.

Saito, M. (1998). *Job and workplace adaptation engineering-human perception-action coupling*. Tokyo, Japan: Nihon Shuppan Service.

Senge, P. (1990). *The fifth discipline: The art and science of learning organization*. New York, NY: Currency Doubleday.

CHAPTER 8

ALLOCATION OF NURSING WORKFORCE IN HOSPITAL WARDS WITH MIXED PATIENT NEEDS

Issachar Gilad

Faculty of Industrial Engineering and Management
Technion-Israel Institute of Technology

Ohad Khabia

Faculty of Industrial Engineering and Management
Technion-Israel Institute of Technology

OVERVIEW

This chapter introduces a procedure for allocating a quality-driven nursing workforce in a hospital's inpatient wards for the best treatment. The procedure is designed to offer a systematic approach for classifying the patient mix, assigning performance times for nursing activities, and forecasting the ward patient mix, taking into consideration the preferred nursing care policy. Unlike the scheduling of nurses done in the health industry today, which is basically intuitive, we propose a scientific approach to the question, What is the best number of nurses for a given mix of patients? The proposed methodology has been implemented in three hospital inpatient wards in a fairly large medical center in central Israel. Findings indicated that this scientific approach allows optimization of nursing allocation in hospital inpatient wards. In some cases, the current allocation patterns were found to be adequate whereas in other cases understaffing was revealed. As a result, staffing levels can be calculated according to the desired quality level. From a practical viewpoint, the proposed procedure enables policy makers to choose the best collection of nurses to meet a desirable level of the ward's nursing treatment quality.

INTRODUCTION

The study of the allocation of nurses for a given mix of patients addresses a frequent question asked by hospital management who are trying to provide a scientific answer to the number of nurses to assign in a given shift on a given day in a hospital's inpatient ward. What is the best economical number of nurses that will provide quality service to patients with different nursing needs? To tackle this question, a procedure for allocating a full-time equivalent (FTE) nursing workforce in hospitals' inpatient wards is introduced. The search for appropriate answers leads us to develop a methodology to meet the demands of service in a dynamic mix of patients by calculating the appropriate number of nurses based on performance assessment and forecasting. The study is of special interest to decision makers and legislators at the regional and national level who are concerned by the shortage in nursing staff, as forecasted by the Bureau of Health Professions (2002).

The volume of the allocated nursing staff in a hospital ward is found to directly influence patients' outcomes. Needleman, Buerhaus, Mattke, Stewart, and Zelevinsky (2002) found an association between the proportion of the total hours of nursing care provided by registered nurses and six outcomes among medical patients (length of stay, rates of urinary tract infections, upper gastrointestinal bleeding, hospital-acquired pneumonia, shock or cardiac arrest, and failure to rescue).

Dealing with the question of how to calculate the appropriate number of nurses in a given shift we assess that the number of nurses relies on the following five parameters:

1. The number of patients in a ward on a given day
2. Patients' nursing severity
3. The nurses' level of expertise available in the ward at a given time
4. Nursing treatment policy
5. The desirable level of treatment quality

To determine correct values for these parameters a methodology is suggested. The methodology consists of the following:

1. Forecasting the patient mix that will require nursing care on any given future day by a data-mining mechanism

2. Establishing performance times for a list of key direct and indirect activities as performed by nurses in a given hospital establishment
3. Calculating and assigning quality measures of nursing performance
4. Developing an "intelligent allocation procedure" for assigning the best nursing workforce for a given patient mix

This is based on a predetermined policy for nursing treatment and on a desirable level of a ward's nursing treatment quality.

The methodology has been implemented in three hospital wards in a fairly large medical center in central Israel, which has about 1,500 beds. The participating wards were two internal wards, which specialize in tropical and joint illnesses, and one surgical ward, which specializes in cancer diseases. After data from the three wards was collected, its analysis allowed us to compare current allocation patterns to the recommendations proposed by the methodology. In some cases the current allocation patterns were found to be adequate whereas in other cases understaffing patterns were revealed. For example, the study showed that an average of 1.1 FTE nurses should be added to the four FTE nurses normally allocated in the surgical unit in the evening shifts in order to meet patient treatment needs. Furthermore, the method will enable policymakers to choose the desirable level of the ward's nursing treatment quality. In return, the appropriate nursing size that needs to be allocated will be calculated.

METHOD

The method supports a procedure to allocate the most economic nursing staff to provide the desired level of patient treatment in a ward. The procedure's main stages are seen in Figure 8.1.

Patient Classification

One of the most critical and continuing concerns facing administrations of health care delivery systems is the management and effective application of a realistic nurse staffing system based on patient acuity levels. The number of nursing personnel needed in a nursing ward can be properly determined only by a system that effectively evaluates and assesses patient needs concurrently with the individual capabilities of the nursing staff assigned to the ward.

The U.S. Department of Health published its system for patient classification (Department of Health and Human Services, 1987). This system was adopted by health administrators as an "ideal" classification procedure. Such a patient classification procedure enabled nursing staff members to classify patients according to the intensity that the nursing care required. The five patient classification levels were minimal, partial, full, intense, and special.

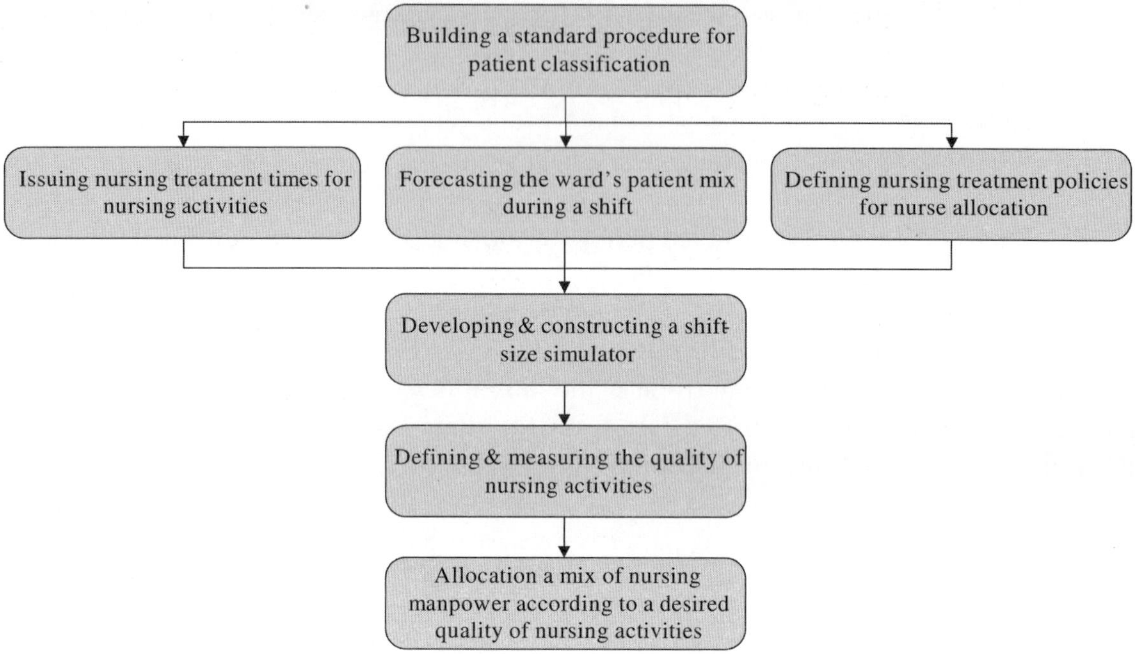

Figure 8.1. Nursing staff allocation procedure for hospital inpatient wards.

A patient whose nursing care needs are entirely met by routine ward activities is to be classified as level 1—the minimal nursing care category. This category of patients is characterized by (a) mild symptoms, usually considered in the later stages of convalescence; (b) little or no deviation from normal behavior patterns; (c) no restriction of activity; (d) simple treatments and few medications; and (e) follow-up instructions and supervised practice in self-care measures in preparation for discharge.

The other range of the patient classification is represented by level 5—the special nursing care category. This category of patients is characterized by (a) a need for life-saving measures promptly and constantly; (b) an extreme deviation from normal behavior patterns; (c) a need for special nursing activities such as respiration, frequent lab tests, and frequent blood infusions; and (d) a need for continuous treatment and observation frequently in an isolation room.

Nursing Treatment for Nursing Activities

The number of FTE nurses required to meet patients' nursing needs in a shift is directly linked to the volume of the nursing work content. Nursing work content is the total amount of time it takes to complete the activities that nurses need in order to execute the care required by ward patients.

This study used a time and motion technique to observe the wards' registered nurses and measure the duration of their nursing activities. Along with measuring the activity duration, additional information was documented in the data collection sheet. This information referred to the patient's classification level and to the status of the patient in the shift being investigated. Patients' statuses were classified as belonging to one of the following three groups:

1. *Admitted patient.* A patient who was hospitalized during the observed shift, having come to the ward from the emergency room or as a result of being transferred from another ward;
2. *Discharged patient.* A patient who was released during the observed shift from the ward to the community (home or institute) or was transferred to another ward; and
3. *"Veteran" patient.* A patient who was not classified in the previous groups and whose condition was classified as belonging to one of the five nursing levels.

All nursing activities were classified as either direct or indirect nursing activities:

1. *Direct nursing activities.* Direct nursing activities are the activities performed in the presence of the assigned patient or family including all preparations and assessments that support these activities such as assessing patient needs; administration of medications; all treatments and procedures; obtaining specimens; and all aspects of the basic physical care associated with bathing and grooming, eating, toileting, and ambulation. This also includes explanations given to the patient and family regarding the mentioned activities.
2. *Indirect nursing activities.* Indirect nursing activities are the activities that are not related to caring for the assigned patient, such as giving or receiving an updated description of the patient's status for the purpose of shift-to-shift continuity; clerical work; cleaning; checking; and looking for equipment, personal needs, and fatigue allowances. The measured time of direct nursing activities was derived in order to produce time distributions for direct nursing care to all patients of every status. The measured time of the indirect nursing activities was derived in order to produce shift timeframes for the nursing staff.

Forecasting Ward Patient Mix

Forecasting the ward patient mix in a shift is a structural procedure. A procedure considered in 10 steps is shown in Figure 8.2. In the upper part of each rectangle, a short description of the step is presented, while in the lower part a short description of the step's outcome appears (a forecast for 2007 is given as an example).

The patient mix in an 8-hour shift is dynamic and is composed of three general patient "populations" that differ from each other in their volume and in the treatment they receive. These populations (or statuses as termed previously) are admitted patients (A_s), discharged patients (D_s), and veteran patients (V_s). The data, which are stored in the Health Care Organization's (HCO) databases, vary from one organization to another. In the investigated HCO, the following relevant data was available: admissions per shift (A_s), discharges per day (D), and the midnight census (MC). Table 8.1 demonstrates how to forecast future populations in a shift (A_s, D_s, and V_s) using the existing available data. N_s^{Beg} and N_s^{End} refer to the number of patients at the beginning and end of a shift, respectively.

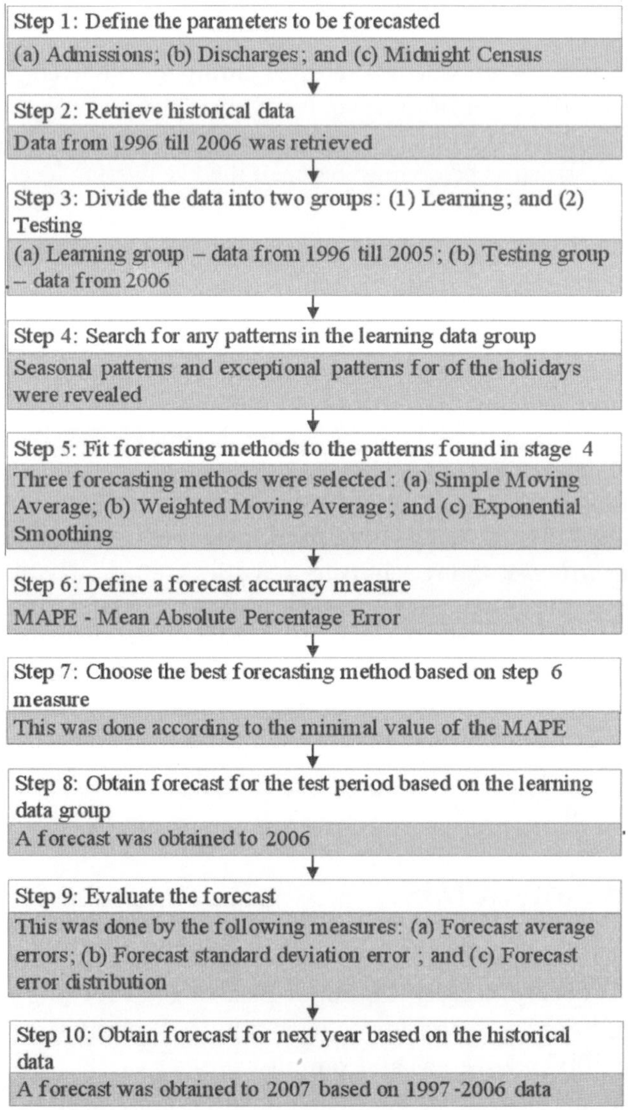

Figure 8.2. The 10 steps of the forecasting procedure.

Table 8.1. Relation Between Admissions per Shift, Discharges per Day, and Midnight Census to Three Patient Populations in a Shift—Admitted Patients, Discharged Patients, and "Veteran" Patients

	Day (7:00–15:00)	Evening (15:00–23:00)	Night (23:00–7:00)
N_s^{beg}	$MC + 7/8 \cdot A_{nigt}$	N_{day}^{end}	N_{eve}^{end}
A_s	A_{day}	A_{eve}	A_{nig}
D_s	$0.35D$	$0.65D$	—
V_s	$N_{day}^{beg} - D_{day}$	$N_{eve}^{beg} - D_{eve}$	N_{nig}^{beg}
N_s^{end}	$N_{day}^{beg} + A_{day} - D_{day}$	$N_{eve}^{beg} + A_{eve} - D_{eve}$	$N_{nig}^{beg} + A_{nig}$
Remark	On weekends and holidays: $D_{day} = 0.80D$	On weekends and holidays: $D_{eve} = 0.20D$	

Nursing Treatment Policies for Nurse Allocation

Nursing treatment policies determine the allocation of nursing staff to the patient mix. Nursing personnel are distinguishable by their individual expertise and responsibilities. The staff was divided into three groups: (a) nurse (N)—nurses hold a current license to practice within the scope of professional nursing; (b) expert nurse (EN)—these nurses are characterized by high nursing skills, motivation, and relatively wide nursing knowledge; and (c) head nurse (HN)—expert nurse who are administratively responsible for a designated hospital ward on an 8-hour basis.

Five nursing care policies are proposed and demonstrated in Figure 8.3. In policy A the staff is divided only into nurses and a head nurse who are assigned to the entire patient mix. Policies B and D suggest assigning nurses to nursing levels 1 to 3 patients, and the head nurse and expert nurses to nursing levels 4 to 5 patients. The difference between nursing policies B and D is in the weekday day shifts. Whereas policy B suggests that the head nurse devotes part of his or her shift (e.g., 10%) to nursing care, policy D proposes that the head nurse concentrates only on his or her administrative responsibilities and not be involved in the operation of caregiving activities. Policies C and E suggest assigning the head nurse and nurses to nursing levels 1 to 3 patients, while expert nurses are assigned to nursing levels 4 and 5 patients. The difference between nursing policies C and E is similar to the difference between policies B and D, respectively.

Shift-Size Simulator

The use of a simulator to support the decision tool was based on the following. First, a tool was required to store and retrieve data, part of which was derived from statistical distributions.

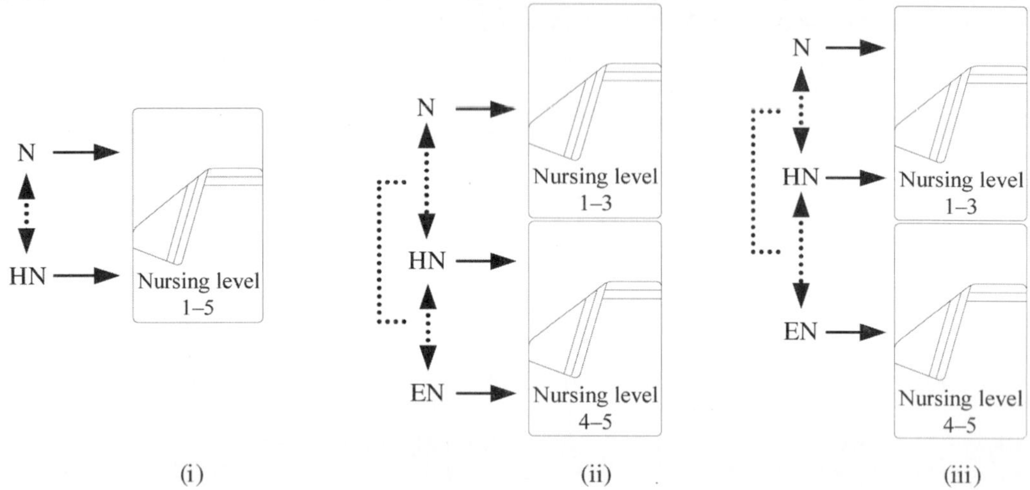

Figure 8.3. Nursing care policies (i) A, (ii) B and D, and (iii) C and E.

Second, a simulator helps give the user a friendly graphics user interface (GUI) for inserting input parameters. Third, it solved the issue of how to give the user recommendations based on data, input parameters, and algorithms.

The simulator was developed using MATLAB code. The data stored in the simulator were direct time distributions, nursing staff shift timeframes, and forecasts of the patient mix. The user inserts into the GUI the starting day and duration of the desired planning period and the chosen nursing care policy. The algorithm by which the recommendations are calculated applies to each nursing care policy. These recommendations are on a shift basis and include FTE for the nursing staff, total direct caregiving time, and total patient forecasts.

Quality of Nursing Activities

A set of direct nursing activities was selected in order to define nursing treatment quality. This was carried out in two main steps. Step 1 converts the result of the direct nursing activity to a quality score:

$$Q_i = \begin{cases} R_i & R_i \geq M_i \\ 0 & \text{Otherwise,} \end{cases} \qquad (1)$$

where R_i is the percentage of patients for whom the direct nursing activity i was carried out (derives from the HCO database) and M_i is the threshold score value for the direct nursing activity i. Usually M_i ([70,80].

Step 2 deals with computing the weighted average score for the nursing treatment quality:

$$Q = \Sigma_i Q_i \cdot W_i, \qquad (2)$$

where W_i is the relative weight for quality of direct activity *i*. This is determined by using the paired comparison technique (Saaty, 2000).

Allocating the Best Mix of Nursing Personnel

The final stage of the methodology deals with formalizing the link between allocating FTE nursing personnel in order to achieve the targeted nursing treatment quality in a ward. The duration of direct nursing treatment activities for a given patient population provides random variables. They are undependable and derived from the same statistical distribution with the same mean (μ_i) and the same standard deviation (σ_i). These preliminary conditions allow us to use the normal approximation of the central limit law. Using this approximation means that the sum of random variables is normally distributed. Hence it becomes possible to use the laws of probability. The probability that the sum of the duration of all direct nursing treatment activities ($\sum_i^n x_i$) will be equal or less than the nurses' availability (A_F) is $1 - \alpha$ as stated in the following equation:

$$P\left(\sum_i^n x_i \le A_F\right) = 1 - \alpha \qquad (3)$$

The nursing treatment quality score (Q_F) is given in the following equation:

$$Q_F = Min\left\{Q_I \cdot \left\{1 + \frac{A_F - A_I}{A_I} \cdot P_Q\right\}, 100\%\right\}, \qquad (4)$$

where Q_I is the treatment quality score in the initial state deriving from allocating nurses according to average treatment times, A_I is the nurses' availability according to average treatment times, and P_Q is the proportion of the direct treatment time associated with performing quality activities out of the total direct nursing time. The assumption behind this calculation is that the quality of nursing treatment will increase if the nurses are more available to perform nursing activities that include quality activities.

FINDINGS: CASE STUDY IN HOSPITAL'S INTERNAL WARD

The methodology was implemented in three hospital wards in a fairly large medical center in central Israel, which holds about 1,500 beds. The wards were two internal wards specializing in tropical and joint illnesses and one surgical ward specializing in cancer diseases. Here we provide a good demonstration of the methodology, as used in internal ward C.

Historical patient classification data revealed that the majority of patients were classified as belonging to one of the first three nursing levels. The patient mix in internal ward C for nursing levels 1 to 5 was 19%, 43%, 28%, 6%, and 4%, respectively.

A total of 19 full shifts were observed using the time and motion technique. Eight nurses were observed in the day shifts, and six nurses and five head nurses were observed in the evening shifts. Direct and indirect nursing activities were measured from the time and motion studies to assist in deriving direct time distributions and shift timeframes. Time distributions, mean, standard deviation, minimum value, and maximum value for direct nursing care for internal ward C are given in Table 8.2. Indirect nursing activities were measured, since they are a nonneglectable part of nursing daily activities. These activities were derived from eight time and six motion studies carried out on nurses and expert nurses in the day and evening shifts, respectively, and from five studies carried out on head nurses in the evening shift. Personal needs and fatigue allowances were calculated from the International Labour Office recommendations for allowances (ILO, 1979). A total of 14% allowances were summed after assigning 5% to personal allowance, 4% to basic fatigue allowance, 2% to standing and to close attention allowances, and 1% to mental strain allowance.

The shift timeframe for nurses in the morning shift, for example, was equal to the shift duration (480 minutes) minus the total indirect time (140.1 minutes).

To allocate the FTE nursing based on the forecasted patient mix, it was necessary to know the rates of patient mix. As explained earlier, midnight census, discharges, and admissions are the basic parameters for forecasting the patient mix in a shift.

Forecasted midnight census rates for the year of 2007 in internal ward C are given in Table 8.3. These occupancy rates were based on learning the patterns revealed in the midnight census data from 1996 to 2005.

The ward's quality of nursing treatment score was derived from a set of direct nursing activities. These activities and their relative weights are given in Table 8.4.

An 8-week pilot study was conducted in internal ward C. The results, which are given in Table 8.5, contain the following parameters: (a) weighted average nursing treatment quality score; (b) total FTE nursing staff that were assigned by the head nurses in the current state and the average number of nurses per shift; (c) total FTE nursing staff that the simulator recommended, using nursing care policy E and the average number of nurses per shift; and (d) the difference between the actual FTE and the recommended FTE allocation.

Figure 8.4 illustrates the correlation between the nursing service levels and the number of FTE nurses per shift and the nursing treatment quality score. As stated earlier, increasing the service level $(1 - \alpha)$ will increase the corresponding percentage $(z_{1-\alpha})$, which will increase the nurses' availability to provide a treatment for a mix of patients. A higher value for the nurses' availability translates directly into a greater number of FTE nurses per shift.

Table 8.2. Time Distributions, Mean, Standard Deviation, Minimum Value, and Maximum Value for Direct Nursing Care for Classified Veteran Patients, Admissions, and Discharges in the Day and Evening Shifts

Shift	*Patient status*	*Distribution*	*M*	*SD*	*Minimum value*	*Maximum value*
Day	1	12+18 beta (0.6, 0.8)	19.5	6.6	12.6	29.3
	2	12+30 beta (0.7, 0.9)	24.9	9.4	12.3	41.7
	3	13+ weibull (21.9, 1.9)	33.9	16.0	13.6	65.6
	4	24+ gamma (39.6, 0.7)	52.1	27.2	24.3	97.9
	5	84+54 beta (0.2, 0.3)	101.0	24.9	84.1	138.0
	Admissions	15+ exponential (15.8)	30.8	13.4	15.2	59.7
	Discharges	13+11 beta (0.7, 0.8)	18.3	4.2	13.3	23.6
Evening	1	Triangular (7, 15, 37)	19.7	7.9	7.3	36.3
	2	Normal (22.4, 7.5)	22.4	7.5	7.4	39.9
	3	16+23 beta (0.8, 1.1)	26.1	6.7	16.2	38.1
	4	Normal (38.3, 17.8)	38.3	17.8	13.5	81.0
	5	Uniform (23, 93)	56.1	28.5	23.1	92.9
	Admissions	15+ exponential (15.8)	30.8	13.4	15.2	59.7
	Discharges	Normal (9.7, 5.1)	9.7	5.1	2.4	21.8

Table 8.3. Midnight Census Rates, Year of 2007

	Jan	*Feb*	*Mar*	*Apr*	*May*	*Jun*	*Jul*	*Aug*	*Sep*	*Oct*	*Nov*	*Dec*
Sunday	43.8	42.5	40.3	38.5	42.9	39.0	42.6	38.1	43.9	39.1	40.2	42.2
Monday	42.6	39.5	37.9	38.7	43.0	39.3	41.3	38.1	42.6	39.0	39.4	42.2
Tuesday	41.4	39.5	38.8	40.7	41.9	38.6	39.6	38.1	42.6	38.3	38.8	39.5
Wednesday	41.0	40.2	38.3	39.0	38.9	37.2	39.7	36.3	39.1	41.0	41.1	40.5
Thursday	40.3	37.7	36.5	35.6	35.7	36.4	36.2	34.4	36.1	43.3	38.4	38.8
Friday	40.0	36.8	36.0	36.8	37.1	34.0	34.8	33.9	36.3	38.8	35.8	38.3
Saturday	45.9	41.5	41.1	40.4	42.3	38.8	42.8	37.1	43.3	43.5	40.8	44.1

Table 8.4. Quality Nursing Activities and Their Relative Weights

#	*Quality nursing activity*	*Wi (%)*
1	Measuring vital signs	11
2	Wearing identification bracelet	
3	Patient briefing in admission	10
4	Norton scale implementation	
5	Pain scale implementation	
6	Estimation of falling risks	
7	Nursing anamnesis implementation	
8	Daily pain scale implementation	
9	Patient classification implementation	
10	Lansky scale implementation	8

Table 8.5. Comparison Between Current and Simulator FTE Nursing Staff Allocation for the Day and Evening Shifts

Shift	*Quality score*	*Current allocation*		*Simulator allocation*		*Difference*		
		Total	*Per shift*	*Total*	*Per shift*	*Total*	*Per shift*	*%*
Day	93.2	269.5	4.8	294.0	5.3	24.5	0.5	9.1
Evening		200.6	3.6	222.0	4.1	31.0	0.5	15.4

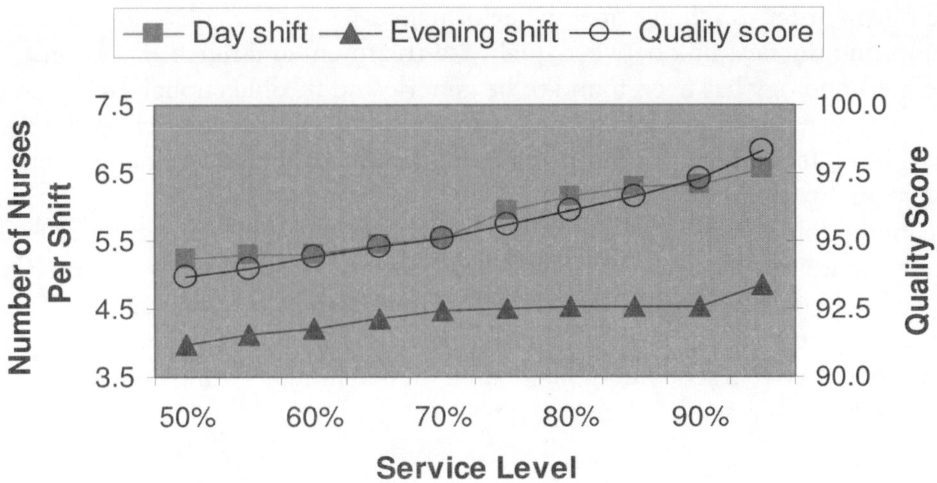

Figure 8.4. Nursing service level versus average number of nurses per shift and nursing treatment quality score.

DISCUSSION

This chapter presented a procedure for defining the best mix of FTE nurses to be allocated in order to provide good care services to a given variety of patients in a hospital inpatient ward. The authors discussed how to relate nursing services to the quality of performance in a straightforward way. These address a very critical issue in a world where a shortage of nurses in the years to come is predicted by every public health authority.

A way to forecast the patient mix in a given shift is introduced. This was done in two steps. Step 1 focused on forecasting the parameters of ward midnight censuses (Table 8.3) and patient admissions and discharges based on the method presented in Figure 8.2. Step 2 focused on formalizing the ratios among the patient mix forecasting parameters in a shift as presented in Table 8.1. Thereafter the volumes of three patient populations were derived for a given shift: veteran, admitted, and discharged patients. The case study presented here validates the theory in a functioning hospital. The study findings relating to weeks 43 to 50 of 2007 are presented in Table 8.5. The findings presented in Figure 8.4 shows as well the formulized correlation between the quality of nursing treatment and the number of nurses.

In contrast to present staff allocation in hospitals, which is based on nonschematic rules of thumb—intuitive by nature, the present study has proposed a schematic, scientific approach to the question of what is the best number of nurses for a given mix of patients. The methodology presented in this chapter has, therefore, successfully addressed a problem hospital administrators all over the world grapple with daily: how to allocate nurses in a scientific way in order to achieve quality performance. Moreover, it has done even more; the method can find the best link between quality of treatment and nurse allocation.

The methodology has been found to be generic and flexible enough to be used by head nurses. Our experience showed that the procedure has been most appreciated by the nursing managerial level. Its accuracy has been achieved due to a systematic modeling of the nursing workload in an inpatient ward.

The methodology presented can be used as a basis for future practical studies. It will be useful to integrate this tool into the HCO information system. In this way the forecasting procedure should proceed with minimum effort in a straightforward manner. It would also be beneficial to integrate this tool into an automated roster tool. This way nursing managers would be able to derive a scheduling plan based on dynamic FTE nursing patterns that rely on the methodology presented in this chapter.

REFERENCES

Bureau of Health Professions. (2002). Projected supply: Demand and shortages of registered nurses: 2000–2020. Retrieved February 20, 2008, from ftp://ftp.hrsa.gov/bhpr/workforce/rnsupplyanddemand2002.pdf

Department of Health and Human Services. (1987). Patient classification and nurse staffing system (Indian Health Service Circular No. 87-1). Retrieved February 20, 2008, from http://www.ihs.gov/PublicInfo/Publications/IHSManual/Circulars/Circ87/Circ87_01/circ87_01.htm

International Labour Office (ILO). (1979). *Introduction to work study* (3rd ed.). Geneva, Switzerland: International Labour Office.

Needleman, J., Buerhaus, P., Mattke, S., Stewart, M., & Zelevinsky., K. (2002). Nurse-staffing levels and the quality of care in hospitals. *New England Journal of Medicine*, *346*(22), 1715–1723.

Saaty, T. A. (2000). *Fundamentals of decision making and priority theory* (2nd ed.). Pittsburgh, PA: RWS Publications.

PART V

TECHNIQUES

CHAPTER 9

SELECTION OF THE BEST MODEL FITTED BY LINEAR REGRESSION WITH A MULTIATTRIBUTE FOCUS

Jorge Luis García A.

Department of Industrial and Manufacturing Engineering
Autonomous University of Ciudad Juárez

Salvador A. Noriega M.

Department of Industrial and Manufacturing Engineering
Autonomous University of Ciudad Juárez

OVERVIEW

Generally, the performance of too many industrial processes can be explained by several factors. Researchers develop models, such as regression, to explain the largest part of the process variability with the least error possible. In some cases, the variability can be explained by too many variables and several models can be developed with subsets of those variables, appearing as a problem for the model selection. In order to aid the selection of the best model, several parameters can be applied, such as the mean square error, the coefficients of multiple determination or adjusted determination, and the Mallows C_p index, sometimes proposing different solutions to the same problem, which can be a cause for confusion. Therefore, this chapter proposes a multiattribute model based in Technique for Order Preference by Similarity to an Ideal Solution (TOPSIS); the model logic considers all the models like alternatives in an n-dimensional space and the parameters as attributes of the alternatives in a k-dimensional space. The model proposed gives the order of preference for the models compared, based in the weighted distance to an ideal model and to another anti-ideal. The model is relatively easy to apply and its application does not require advanced computer skills.

INTRODUCTION

It is generally accepted that the quickest and more economical way to model a variable behavior (Y) is by the use of a single predictor variable with a linear regression model, adjusted by ordinary least squares (OLS). Nonetheless, the common behavior of Y is complex and it is hard to explain it efficiently by just one input variable (Lam & Veall, 2002). For example, let us say that milk sales depend, among other variables, on the population feeding habits, marketing, climate, and points of sale. Therefore, a sales forecast model should include more than one predictor variable of the multiple regression kind (Blyth & Kaka, 2006).

Although, in the vast majority of practical problems, the analyst has a set of candidate regressors on independent variables that should include all of the contributing factors and then he has to define the proper subset that is to be contained in the model. In this fashion, an adjusted regression model is obtained for every subset of regressors, and then the selection problem of the best subset appears (Montgomery, Peck, & Vining, 2007; Preminger, Ben-Zion, & Wettstein, 2007). Nonetheless, the construction of a regression model that only includes a subset of regressors implies some contradicting objectives. The main contradiction is as follows: It is desirable that the model contain as many regressors as possible, assuming that more information might produce a better prediction of Y, but at the same time it is desirable that the model contain the least quantity of regressors, given that the variance of the prediction increases with that quantity (Allen, 1971, 1974; Montgomery et al., 2007). The selection of the best regression equation implies a balance between those two objectives, because the analyst wants models with few variables that contain much explanatory power (Box & Wets, 1973).

The selection of the best regression equation is managed by a trial-and-error process. In fact, there is no one best equation; several of them are good. One has to consider that the algorithms for the variable selection depend so much in the procedure employed and that the annalist might ponder erroneously the data and select a model that is not as good as another. That is why expertise is important in the variable selection and for the determination of the adequate model (Gunts & Mason, 1979).

In the literature, several methods are used for the production of acceptable models, among them the step-by-step regression method and the forward and backward regression method. Those methods differ, essentially, in the way iterations begin in the solutions they provide, frequently distinct. There are several research reports covering the selection of variables with distinct methodologies, such as the ones proposed by Hocking (1972, 1976), Hocking and LaMotte (1973), Cox and Snell (1974), Thompson (1987a, 1987b), and Myers

(1990). Several researchers have investigated the properties of the $\hat{\beta}_p$ y $\hat{\sigma}_p$ estimators of the subsets (Hocking, 1974, 1976; Narula & Rambers, 1972; Rao, 1971; Rosemberg & Levy, 1972; Walls & Weeks, 1969). Also, one has to consider that the commonly used parameters in the selection of the regression model are based in residual mean squares, the mean square error, the coefficient of multiple determination or adjusted determination, and the Mallows C_p index, and every one of them may lead to different solution models. Therefore, the annalist may be a set of regression models with different variables and distinct parameters for a single problem. That is why the authors of this chapter propose a model for the selection of one of those models.

The model being proposed is based on the TOPSIS. In this proposed model, the parameters obtained are considered as attributes and the models are evaluated by means of the TOPSIS. The model's logic resides in the intuitive idea that there must be an ideal model with the least quantity of variables, the minimum mean square error, the minimum mean square, minimum C_p Mallows index, the highest coefficients of multiples, and the adjusted determination.

The TOPSIS is a relatively new and easy-to-apply technique when there is set of alternatives (Hwang & Yoon, 1985; Lai, Liu, & Hwang, 1994). It has been successfully applied in a wide variety of selection problems, such as the recruiting and management of personnel (Chen & Tzeng, 2004), the selection of design and prototypes (Kwong & Tam, 2002), the selection of transport systems (Janic, 2003), the evaluation of manufacturing processes (Milani, Shanian, & Madoliat, 2005), the selection of water distribution systems (Srdjevic, Medeiros, & Faria, 2004), the simulation of multiple responses in stochastic systems and quality control (Yang & Chou, 2005), and the location of industrial plants (Yoon & Hwang, 1995).

METHODOLOGY

The Data

The model proposed is based on the data reported by Hald (1952), where the response variable (y) is the heat in cal/gr of cement produced, which is a function of four mixture components: $X1$ stands for tricalcium aluminate, $X2$ stands for tricalcium silicate, $X3$ stands for tetracalcium aluminumferrite, and $X4$ stands for bicalcium silicate. Concentrations and heat of the cement produced are shown in Table 9.1.

Fifteen models were developed: four of them included one variable; six models contained two variables; four models were composed of three variables; and one model had four variables. The 15 models constitute the alternatives.

Table 9.2 presents the parameters calculated for the models obtained with data given in Table 9.1 and represents the attributes for the evaluation. The first column indicates the name for the parameter estimated, the second column gives the equation for its estimation, and the third column indicates the desirable value in those parameters.

Table 9.1. Cement Data

Observation number	Y_i	X_1	X_2	X_3	X_4
1	78.5	7	26	6	60
2	74.3	1	29	15	52
3	104.3	11	56	8	20
4	87.6	11	31	8	47
5	95.9	7	52	6	33
6	109.2	11	55	9	22
7	102.7	3	71	17	6
8	72.5	1	31	22	44
9	93.1	2	54	18	22
10	115.9	21	47	4	26
11	83.8	1	40	23	34
12	113.3	11	66	9	12
13	109.4	10	68	8	12

Source: Hald (1952).

Parameters and Its Equations

Table 9.2 lists the main formulas used to obtain the values for the parameters that were used as attributes: y is the matrix of the response variable values, x is the matrix of the data obtained in the k variables, $\hat{\beta}$ is the matrix of the estimated coefficients, and ε is the error between a preestablished value and the response variable. The 15 models were adjusted accordingly to the ordinary least squares technique (OLS), using the software named NCSS. Also, it is important to say that like a parsimony principle, the model must have only three parameters, two for variables and one for origin ordinate.

In same sense, SS_T is the regression sum of squares, $SS_R(p)$ is the model sum of squares, n is the quantity of observations, p is quantity of parameters of the models and $\hat{\sigma}^2$ is an estimator of the variance or the mean square error.

The formulas that appear in Table 9.2 are usually used to determine parameters of efficiency for regression models.

$$y = \begin{bmatrix} y_0 \\ y_1 \\ \cdot \\ \cdot \\ y_n \end{bmatrix} \quad (1)$$

Table 9.2. Formulas and Parameters of the Models

Parameter or attribute	Formula	*Nominal value desired*
Square sum of residuals	$SS_{Res} = \varepsilon'\varepsilon = y'y - \hat{\beta}'X'y$	Minimum
Multiple determination coefficient	$R_p^2 = 1 - \dfrac{SS_{Res}(p)}{SS_T(p)} = \dfrac{SS_R(p)}{SS_T}$	Maximum
Adjusted coefficient of determination	$R^2_{Adj,p} = 1 - \left(\dfrac{n-1}{n-p}\right)\left(1 - R_p^2\right)$	Maximum
Square mean error	$MS_{Res} = \dfrac{SS_{Res}(p)}{n-p}$	Minimum
Mallows C_p index	$C_p = \dfrac{SS_{Res}(p)}{\hat{\sigma}^2} - n + 2p$	Minimum
Quantity of parameters	$\hat{\beta} = (X'X)^{-1} X'y$	3, minimum*

* It is supposed that the ideal model must contain three parameters, two of them related to the coefficients of regressors, the other one related to the origin ordinate.

$$X = \begin{bmatrix} 1 & x_{11} & x_{12} & . & x_{1k} \\ 1 & x_{21} & x_{22} & . & x_{2k} \\ . & . & . & . & . \\ . & . & . & . & . \\ 1 & x_{n1} & x_{n2} & . & x_{nk} \end{bmatrix} \quad (2)$$

$$SS_T(p) = y'y - \frac{\left(\sum_{i=1}^{n} y_i\right)^2}{n} \quad (3)$$

$$SS_R(p) = \hat{\beta}'X'y - \frac{\left(\sum_{i=1}^{n} y_i\right)^2}{n} \qquad (4)$$

$$\hat{\beta} = (X'X)^{-1}X'y \qquad (5)$$

The TOPSIS for Model Selection

For the selection of the best model, the TOPSIS was applied. The TOPSIS evaluates alternatives considering them as points in an n-dimensional Euclidean space—the best alternative is the one that has the shortest distance to an ideal model and the largest distance to an anti-ideal model. These two points represent the best and worst nominal values in the attributes (parameters in this case). In the following paragraphs, the alternatives will be represented as $A_1, A_2, \ldots A_{15}$.

Given that an A^k model is considered as a vector in the Euclidean space, it can be represented by means of equation 6, according to Yoon (1980). Also, because every model has one point in the n-dimensional space, the nth attribute can be considered as a vector in the k-dimensional space given by equation 7 ($k = 15$ in this example because there are 15 models in evaluation).

$$A^k = (x_1^k. \ldots x_n^k) \text{ for } k = 1, 2, \ldots 15, \text{ and } n = 1, 2, \ldots 6 \qquad (6)$$

$$X_n = (x_1^k. \ldots x_n^k) \text{ for } k = 1, 2, \ldots 15, \text{ and } n = 1, 2 \ldots 6 \qquad (7)$$

The TOPSIS begins from the assumption that it is possible to create a model with the best values in the attributes. It is called the *ideal model* and can be represented by equation 8 and represents another alternative, usually represented by A^+. In the same sense, then, it is also possible to create a model composed with the worst values. It is named the *anti-ideal model* and can be represented by equation 9. It is important to say that the ideal and anti-ideal models do not exist; they are created with the best and worst values and are used only as a reference in the Euclidean space.

$$A^+ = (x_1^+, x_2^+, \ldots x_n^+) \text{ for } n = 1, 2 \ldots 6 \qquad (8)$$

$$A^- = (\bar{x}_1, \bar{x}_2, \ldots \bar{x}_n) \text{ for } n = 1, 2 \ldots 6 \qquad (9)$$

So, according to the ideal and anti-ideal models, the model or alternative to be selected is represented as a point in the Euclidean space and must be very near to the ideal model but very far from the anti-ideal model. The details of the TOPSIS are explained in the following paragraphs.

All the X_n vectors representing the attributes must be normalized and transformed to a TX_n, accordingly to equation 10, where $\|X_n\|$ represents the Euclidean norm of the attribute vector. This transformation is needed because the attributes might be measured with distinct measuring units or expressed in different scales. After the normalization process, the attributes will be nondimensional values.

$$TX_n = \frac{X_n}{\|X_n\|} = \left(\frac{x_n^1}{\|X_n\|}, \ldots\ldots \frac{x_n^k}{\|X_n\|} \right) \quad (10)$$

$$\|X_i\| = \sqrt{\sum_{i=1}^{k} X_i^{k^2}} \quad (11)$$

Next, the Euclidean distances between the points representing the ideal model and the alternative models being evaluated are calculated, applying equation 12. Equation 13 is used to estimate the distance between points representing the anti-ideal model and those being evaluated or the alternatives.

$$\rho(A^k, A^+) = \left\| w * (TA^k - TA^+) \right\| = \sqrt{\sum_{n=1}^{N} w_1 * (t_1^k, \ldots\ldots, t_n^+)^2} \quad (12)$$

$$\rho(A^k, A^-) = \left\| w * (TA^k - TA^-) \right\| = \sqrt{\sum_{n=1}^{N} w_1 * (t_1^k, \ldots\ldots, t_n^-)^2} \quad (13)$$

where w is a representation of the weight that the annalist has given to the attribute or parameter being evaluated. This is because the annalist maybe has different preference level for the attributes in evaluation. If a weight factor is not integrated, then the annalist is assuming similar weight for every attribute. Finally, the ordering of the models according to the relative closeness to the ideal and anti-ideal models is obtained. The order is given by the index given by equation 14. The selection criteria of the TOPSIS is the selection of the smallest value in $RC(A^k, A^+)$.

$$RC(A^+, A^-) = \frac{\rho(A^k, A^+)}{\rho(A^k, A^+) + \rho(A^k, A^-)} \tag{14}$$

RESULTS

This section is divided into other subsections with more specific information. The first section is related to the models used as alternatives in the evaluation process, the second section presents an evaluation of the data set with traditional techniques, and the third section is related to the evaluation done using the TOPSIS, based in Euclidean distances to ideal and anti-ideal models.

Models to Be Evaluated

The models obtained by NCSS are shown in Table 9.3. The ideal and anti-ideal models are presented in the bottom rows. The ideal model will be the one with only three parameters, the minimum square sum of residuals, the maximum determination coefficient, the smallest mean square error, the maximum adjusted determination coefficient, and the smallest Mallows C_p index. The anti-ideal model is composed of the opposed values to the ideal solution. Also, the last row illustrates values of w used in this example.

In Table 9.3, the values of the ideal model are shown with a superscripted "+" symbol, while the anti-ideal model values are with a superscripted "–" symbol. As one can see, according to the square sum of residuals (SSRes), the model to be chosen is the one that has all the variables because it has the minimum value, and a model with low values in this parameter is desired. In relation to the determination coefficient (R-squared), high values in this parameter are desired and the model to be selected would be $X_1X_2X_4$ or $X_1X_2X_3X_4$, because those have the biggest value. Low values are better if the decision is based in the mean square error (MSERes), hence the model with $X_1X_2X_4$ would be chosen. If the adjusted determination coefficient is considered, then high values are better and the model to be selected would be $X_1X_2X_4$. Finally, if the C_p Mallows index is considered as a decision index, then the model to be selected is X_1X_2.

According to the findings in the previous paragraph, there are three different models proposed that provide a rather satisfactory solution to the same problem according to a different decision index. The models are those that include X_1, X_2, and X_4; and X_1 and X_2.

Getting Models With Traditional Techniques

In order to compare the solution obtained with the proposed method based in the TOPSIS, several solutions were estimated with other methodologies that appear as routines already integrated in the NCSS. These are illustrated in Table 9.4. Four different methodologies

Table 9.3. Parameters for the Models

Model regressors	Model coefficients	P	SSRes(p)	R-squared	MSERes	R-squared adjusted	C_p
X_1	1	2.0	1265.686	0.533948	115.062	0.49158	202.54
X_2	1	2.0	906.3363	0.666268	82.3942	0.63593	142.48
X_3	1	2.0	1939.400⁻	0.285873⁻	176.309⁻	0.22095⁻	315.14⁻
X_4	1	2.0	883.8669	0.674542	80.3515	0.64495	138.730
X_1X_2	2	3.0	57.9045	0.978678	5.7904	0.97441	2.6782⁺
X_1X_3	2	3.0	1227.072	0.548167	122.707	0.4578	198.09
X_1X_4	2	3.0	74.7621	0.972471	7.4762	0.96697	5.4958
X_2X_3	2	3.0	415.4427	0.847025	41.5442	0.81644	62.437
X_2X_4	2	3.0	868.8801	0.68006	86.8880	0.61607	138.22
X_3X_4	2	3.0	175.738	0.93529	17.5737	0.92235	22.373
$X_1X_2X_3$	3	4.0	48.1106	0.982285	5.3456	0.97638	3.0412
$X_1X_2X_4$	3	4.0	47.9727	0.9823⁺	5.3303⁺	0.97645⁺	3.0182
$X_1X_3X_4$	3	4.0	50.8361	0.981281	5.6484	0.97504	3.4968
$X_2X_3X_4$	3	4.0	73.8145	0.97282	8.2016	0.96376	7.3374
$X_1X_2X_3X_4$	4	5.0	47.8636⁺	0.9823⁺	5.9829	0.97356	5
Ideal model		3.0	47.8636	0.9823	5.3303	0.9764	2.6782
Anti-ideal model		5.0	1939.40	0.2858	176.309	0.2209	315.15
w		0.15	0.4	0.1	0.15	0.15	0.05

or criteria to add or remove variables were used: forward regression, backward regression, step-by-step regression, and minimum *MSE*. The model obtained using forward regression has to include the variables X_1 and X_4; in a backward regression, the model must include X_1 and X_2 (given that X_3 and X_4 were removed); in the step-by-step solution, the model includes X_4 and X_1; and finally, if we use the minimum *MSE* criteria, the model must include the variables X_4, X_1, and X_2.

Again, there are three different models proposed as a solution for the same problem using traditional evaluation criteria. Due to this problem, it is not clear which one should be selected as the better model.

The TOPSIS Evaluation

According to the TOPSIS, the first step when selecting any alternative from a set is to make a normalization process, which lets us have nondimensional values. The TOPSIS evaluation

Table 9.4. Models for Forward, Backward, and Step-by-Step Regression

Iteration	*Action*	*Variable*	*R-squared*	*R(MSE)*
		Forward regression		
0	No change		0	15.04372
1	Adding	X_4	0.674542	8.963902
2	Adding	X_1	0.972471	2.734266
3	No change		0.972471	2.734266
		Backward regression		
0	No change		0.982376	2.446008
1	Removed	X_3	0.982335	2.308745
2	Removed	X_4	0.978678	2.406335
3	No change		0.978678	2.406335
		Step-by-step regression		
0	No change		0	15.04372
1	Adding	X_4	0.674542	8.963902
2	No adding		0.674542	8.963902
3	Adding	X_1	0.972471	2.734266
4	No change		0.972471	2.734266
		Minimum MSE		
0	Unchanged		0.000000	15.04372
1	Added	X_4	0.674542	8.963902
2	Added	X_1	0.972471	2.734266
3	added	X_2	0.982335	2.308745
4	Unchanged		0.982335	2.308745

was done in Excel. Table 9.5 illustrates the estimation process of the Euclidean norm for every attribute. The Euclidean norm for every attribute was estimated using equation 11 and appears in the last row.

Once the Euclidean norm was estimated, then the normalization process was applied. That process appears in Table 9.6, and to obtain the normalized values, equation 10 was applied. Please note that the ideal and anti-ideal models are included in this table in the last two rows.

The next step in the TOPSIS is to weigh the attributes in case the annalist has a different preference level. Table 9.3 in the last row illustrates the weight for every attribute in this case of study; another person may have a different preference. In this case the vector w is represented by 0.15, 0.4, 0.1, 0.15, 0.15, and 0.05, meaning that the decision maker has a preference of 0.15 by the number of parameters in the model, 0.4 for the square sum of the

Table 9.5. Euclidean Norm Estimation Process

Regressors in model	*P*	*SSRes(p)*	*R-squared*	*MSERes*	*R-squared adjusted*	C_p
X_1	4	1601961.050	0.2851	13239.263	0.2417	41022.451
X_2	4	821445.4887	0.4439	6788.8042	0.4044	20300.550
X_3	4	3761272.360	0.0817	31084.863	0.0488	99313.219
X_4	4	781220.6969	0.4550	6456.3636	0.4160	19246.012
X_1X_2	9	3352.9311	0.9578	33.5287	0.9495	7.1728
X_1X_3	9	1505705.693	0.3005	15057.0078	0.2096	39239.6481
X_1X_4	9	5589.3716	0.9457	55.8936	0.9350	30.2038
X_2X_3	9	172592.6370	0.7175	1725.9206	0.6666	3898.3790
X_2X_4	9	754952.6282	0.4625	7549.5245	0.3795	19104.768
X_3X_4	9	30883.8446	0.8748	308.8349	0.8507	500.5511
$X_1X_2X_3$	16	2314.6298	0.9649	28.5754	0.9533	9.2489
$X_1X_2X_4$	16	2301.3799	0.9649	28.4121	0.9535	9.1095
$X_1X_3X_4$	16	2584.3091	0.9629	31.9044	0.9507	12.2276
$X_2X_3X_4$	16	5448.5804	0.9464	67.2662	0.9288	53.8374
$X_1X_2X_3X_4$	25	2290.9242	0.9649	35.7951	0.9478	25.0000
Sum	159.00	9453916.525	10.3284	82491.958	9.8359	242772.38
Square root (Euclidean norm)	12.6095	3074.7222	3.2138	287.2141	3.1362	492.7194

residual, and so on. Note that the sum for attribute weights is one. Table 9.7 illustrates the normalized and weighted values.

Once the values are normalized and weighed, then the next step is to estimate the distance of each model to the ideal and anti-ideal model. Table 9.8 and Table 9.9 illustrate that calculation, which was obtained using equations 12 and 13, respectively. The first calculation is related to the distance to the ideal model and the second calculation to the distance to the anti-ideal model.

Once the distances to ideal and anti-ideal models are obtained, then next operation is to estimate the decision index used by the TOPSIS, which is obtained using equation 14. Results are illustrated in Table 9.10, where the last column indicates the order preferences for the index, which is sorted from low to high values of the TOPSIS decision index.

According to Table 9.10 the better proposed model must include the X_1 and X_2 variables because it is the one that has the smallest distance to the ideal solution and the farthest to the anti-ideal solution.

Table 9.6. Normalization Process

Regressor in model	*P*	*SSRes(p)*	*R-squared*	*MSERes*	*R-squared adjusted*	C_p
X_1	0.15861	0.41164	0.16614	0.40061	0.15674	0.41107
X_2	0.15861	0.29477	0.20732	0.28687	0.20277	0.28917
X_3	0.15861	0.63076	0.08895	0.61386	0.07045	0.63959
X_4	0.15861	0.28746	0.20989	0.27976	0.20565	0.28156
X_1X_2	0.23792	0.01883	0.30452	0.02016	0.31070	0.00544
X_1X_3	0.23792	0.39908	0.17057	0.42723	0.14597	0.40203
X_1X_4	0.23792	0.02432	0.30259	0.02603	0.30832	0.01115
X_2X_3	0.23792	0.13512	0.26356	0.14465	0.26033	0.12672
X_2X_4	0.23792	0.28259	0.21161	0.30252	0.19644	0.28052
X_3X_4	0.23792	0.05716	0.29102	0.06119	0.29410	0.04541
$X_1X_2X_3$	0.31722	0.01565	0.30565	0.01861	0.31132	0.00617
$X_1X_2X_4$	0.31722	0.01560	0.30565	0.01856	0.31135	0.00613
$X_1X_3X_4$	0.31722	0.01653	0.30533	0.01967	0.31090	0.00710
$X_2X_3X_4$	0.31722	0.02401	0.30270	0.02856	0.30730	0.01489
$X_1X_2X_3X_4$	0.39653	0.01557	0.30565	0.02083	0.31042	0.01015
Ideal model	0.23792	0.01557	0.30565	0.01856	0.31133	0.00544
Anti-ideal model	0.39653	0.63076	0.08893	0.61386	0.07044	0.63961

Regarding the ideal model, the proposed solution, the model that has X_1 and X_2, only has 20.97% more of a residual sum of squares than the ideal model, it is 0.3686% lower than the determination coefficient, the square mean error is 7.69% higher, the adjusted determination coefficient is 0.2044% lower, and it has the lowest C_p Mallows index. Also, it is composed by two variables, so it is a simple model, but at the same time it explains the issues well. The second best is the model that has variables X_1 and X_4. The distance to the ideal model is larger than to the model occupying first place, but the distance to the anti-ideal solution is only a few hundredths closer.

CONCLUSION

In this chapter we evaluated, with the use of the TOPSIS, several potential models for the prediction of the dependent variable behavior. A multiattribute focus was followed while evaluating simultaneously different parameters of the models. We also showed that by using several weights, the evaluation can reflect the annalist's preferences. For instance, if it is desired that the decision must be made with just one attribute, all the weight has to be

Table 9.7. Attributes Weighting Process

Regressor in model	*P*	*SSRes(p)*	*R-squared*	*MSERes*	*R-squared adjusted*	C_p
X_1	0.02379	0.16466	0.01661	0.06009	0.02351	0.02055
X_2	0.02379	0.11791	0.02073	0.04303	0.03042	0.01446
X_3	0.02379	0.25230	0.00890	0.09208	0.01057	0.03198
X_4	0.02379	0.11498	0.02099	0.04196	0.03085	0.01408
X_1X_2	0.03569	0.00753	0.03045	0.00302	0.04660	0.00027
X_1X_3	0.03569	0.15963	0.01706	0.06408	0.02190	0.02010
X_1X_4	0.03569	0.00973	0.03026	0.00390	0.04625	0.00056
X_2X_3	0.03569	0.05405	0.02636	0.02170	0.03905	0.00634
X_2X_4	0.03569	0.11304	0.02116	0.04538	0.02947	0.01403
X_3X_4	0.03569	0.02286	0.02910	0.00918	0.04411	0.00227
$X_1X_2X_3$	0.04758	0.00626	0.03056	0.00279	0.04670	0.00031
$X_1X_2X_4$	0.04758	0.00624	0.03057	0.00278	0.04670	0.00031
$X_1X_3X_4$	0.04758	0.00661	0.03053	0.00295	0.04663	0.00035
$X_2X_3X_4$	0.04758	0.00960	0.03027	0.00428	0.04609	0.00074
$X_1X_2X_3X_4$	0.05948	0.00623	0.03057	0.00312	0.04656	0.00051
Ideal model	0.03569	0.00623	0.03057	0.00278	0.04670	0.00027
Anti-ideal model	0.05948	0.25230	0.00889	0.09208	0.01057	0.03198

assigned to that specific attribute, with a weight of one for that attribute and zero for the others. As can be seen in this example, the method is easy to use and gives similar results to the ones obtained with other commonly used selection methods but allows the integration that the annalist prefers and evaluates simultaneously more than one parameter. The method has a noticeable disadvantage: it is only useful for problems that involve few regressors variables, given that the quantity of models to evaluate grows exponentially accordingly to a 2^k function, where k is the number of regressors variables. That is why, for a 10-variable problem, the number of alternative choices is $2^{10} = 2016$, which will require too many calculations and programming time.

REFERENCES

Allen, D. M. (1971). Simultaneous inference and the choice of variables subsets. *Technometrics*, *13*, 469–475.

Allen, D. M. (1974). The relationship between variable selection and data augmentation and a method for prediction. *Technometrics*, *16*, 125–127.

Table 9.8. Distances to the Ideal Model

Model regressors	*P*	*SSRes(p)*	*R-squared*	*MSERes*	*R-squared adjusted*	C_p	*Sum*	*Distance*
X_1	0.00014	0.02510	0.00019	0.00328	0.00054	0.00041	0.02967	0.17225
X_2	0.00014	0.01247	0.00010	0.00162	0.00027	0.00020	0.01480	0.12164
X_3	0.00014	0.06055	0.00047	0.00797	0.00131	0.00101	0.07145	0.26730
X_4	0.00014	0.01183	0.00009	0.00154	0.00025	0.00019	0.01404	0.11848
X_1X_2	0.00000	0.00000	0.00000	0.00000	0.00000	0.00000	0.00000	0.00134
X_1X_3	0.00000	0.02353	0.00018	0.00376	0.00062	0.00039	0.02848	0.16877
X_1X_4	0.00000	0.00001	0.00000	0.00000	0.00000	0.00000	0.00001	0.00373
X_2X_3	0.00000	0.00229	0.00002	0.00036	0.00006	0.00004	0.00276	0.05251
X_2X_4	0.00000	0.01141	0.00009	0.00181	0.00030	0.00019	0.01380	0.11746
X_3X_4	0.00000	0.00028	0.00000	0.00004	0.00001	0.00000	0.00033	0.01818
$X_1X_2X_3$	0.00014	0.00000	0.00000	0.00000	0.00000	0.00000	0.00014	0.01190
$X_1X_2X_4$	0.00014	0.00000	0.00000	0.00000	0.00000	0.00000	0.00014	0.01190
$X_1X_3X_4$	0.00014	0.00000	0.00000	0.00000	0.00000	0.00000	0.00014	0.01190
$X_2X_3X_4$	0.00014	0.00001	0.00000	0.00000	0.00000	0.00000	0.00016	0.01248
$X_1X_2X_3X_4$	0.00057	0.00000	0.00000	0.00000	0.00000	0.00000	0.00057	0.02380

Table 9.9. Distances to the Anti-Ideal Model

Model regressors	*P*	*SSRes(p)*	*R-squared*	*MSERes*	*R-squared adjusted*	C_p	*Sum*	*Distance*
X_1	0.00127	0.00768	0.00006	0.00102	0.00017	0.00013	0.01034	0.10167
X_2	0.00127	0.01806	0.00014	0.00241	0.00039	0.00031	0.02258	0.15027
X_3	0.00127	0.00000	0.00000	0.00000	0.00000	0.00000	0.00127	0.03569
X_4	0.00127	0.01886	0.00015	0.00251	0.00041	0.00032	0.02352	0.15336
X_1X_2	0.00057	0.05991	0.00046	0.00793	0.00130	0.00101	0.07118	0.26679
X_1X_3	0.00057	0.00859	0.00007	0.00078	0.00013	0.00014	0.01027	0.10136
X_1X_4	0.00057	0.05884	0.00046	0.00777	0.00127	0.00099	0.06990	0.26439
X_2X_3	0.00057	0.03931	0.00030	0.00495	0.00081	0.00066	0.04660	0.21587
X_2X_4	0.00057	0.01940	0.00015	0.00218	0.00036	0.00032	0.02297	0.15157
X_3X_4	0.00057	0.05264	0.00041	0.00687	0.00113	0.00088	0.06250	0.25000
$X_1X_2X_3$	0.00014	0.06054	0.00047	0.00797	0.00131	0.00100	0.07143	0.26726
$X_1X_2X_4$	0.00014	0.06055	0.00047	0.00797	0.00131	0.00100	0.07144	0.26728
$X_1X_3X_4$	0.00014	0.06036	0.00047	0.00794	0.00130	0.00100	0.07122	0.26687
$X_2X_3X_4$	0.00014	0.05890	0.00046	0.00771	0.00126	0.00098	0.06945	0.26353
$X_1X_2X_3X_4$	0.00000	0.06055	0.00047	0.00791	0.00130	0.00099	0.07122	0.26688

Table 9.10. Decision Index

Regressors in model	Distance to A+	Distance to A–	Decision index	Order
X_1X_2	0.00133634	0.26679201	0.00498394	1
X_1X_4	0.00372557	0.26438854	0.01389546	2
$X_1X_2X_4$	0.01189583	0.26728302	0.04261008	3
$X_1X_2X_3$	0.01189588	0.26726307	0.04261327	4
$X_1X_3X_4$	0.01190373	0.26686715	0.04270075	5
$X_2X_3X_4$	0.01248326	0.26352941	0.04522712	6
X_3X_4	0.01817815	0.2499961	0.06778484	7
$X_1X_2X_3X_4$	0.02379554	0.266875	0.08186431	8
X_2X_3	0.05251123	0.21586834	0.1956603	9
X_4	0.11848456	0.15336011	0.43585389	10
X_2X_4	0.1174605	0.15156663	0.4366121	11
X_2	0.12164378	0.1502742	0.44735467	12
X_1X_3	0.16876727	0.10135766	0.62477488	13
X_1	0.1722485	0.1016675	0.62883693	14
X_3	0.26729923	0.03568732	0.88221483	15

Blyth, K., & Kaka, A. (2006). A novel multiple linear regression model for forecasting S-curves. *Engineering Construction & Architectural Management*, *13*(1), 82–95.

Box, G. E. P., & Wetz, J. M. (1973). *Criterion for judging the adequacy of estimation by an approximating response polynomial* (Technical Report No. 9). Madison, WI: Department of Statistics, University of Wisconsin.

Chen, M. F., & Tzeng, G. H. (2004). Combining gray relation and TOPSIS concepts for selecting an expatriate host country. *Mathematical and Computer Modeling*, *40*, 1473–1490.

Cox, D. R., & Snell, E. J. (1974). The choice of variables in observational studies. *Applied Statistics*, *23*, 51–59.

Gunts, R. F., & Mason, R. L. (1979). Some considerations in the evaluation of alternative prediction equations. *Technometrics*, *21*, 55–63.

Hald, A. (1952). *Statistical theory with engineering applications*. New York, NY: Wiley.

Hocking, R. R. (1972). Criteria for selection of a subset regression: Which one should be used? *Technometrics*, *14*, 967–970.

Hocking, R. R. (1974). Misspecification in regression. *American Statistics*, *28*, 39–40.

Hocking, R. R. (1976). The analysis and selection of variables in linear regression. *Biometrics*, *32*, 1–49.

Hocking, R. R., & LaMotte, L. R. (1973). Using the SELECT program for choosing subset regressions. In W. O. Thompson & F. B. Cady (Eds.), *Proceedings of the University of Kentucky conference on regression with a large number of predictors variables*. Lexington: University of Kentucky.

Hwang, C. L., & Yoon, K. (1985). Manufacturing plant location analysis by multiple attribute decision making: Part I—single-plant strategy. *International Journal of Production Research*, *23*, 345–359.

Janic, M. (2003). Multicriteria evaluation of high-speed rail, transrapid maglev, and air passenger transport in Europe. *Transportation Planning and Technology*, *26*(6), 491–512.

Kwong, C. K., & Tam, S. M. (2002). Case-based reasoning approach to concurrent design of low power transformers. *Journal of Materials Processing Technology*, *128*, 136–141.

Lai, Y. J., Liu, T. Y., & Hwang, C. L. (1994). TOPSIS for MODM. *European Journal of Operational Research*, *76*(3), 486–500.

Lam, J. P., & Veall, M. R. (2002). Bootstrap prediction intervals for single period regression forecasts. *International Journal of Forecasting*, *18*(1), 125–130.

Milani, A. S., Shanian, A., & Madoliat, R. (2005). The effect of normalization norms in multiple attribute decision-making models: A case study in gear material selection. *Structural Multidisciplinary Optimization*, *29*(4), 312–318.

Montgomery, D. C., Peck, E. A., & Vining, G. G. (2007). *Introduction to Linear Regression Analysis*. Phoenix, AZ: Wiley.

Myers, R. H. (1990). *Classical and modern regression with applications* (2nd ed.). Boston, MA: PWS-Kent.

Narula, S., & Rambers, J. S. (1972). Carta al editor. *American Statistical*, *26*, 42.

Preminger, A., Ben-Zion, U., & Wettstein, D. (2007). The extended switching regression model: Allowing for multiple latent state variables. *Journal of Forecasting*, *26*(7), 457–473.

Rao, P. (1971). Some notes on misspecification in regression. *American statistics*, *25*, 37–39.

Rosemberg, S. H., & Levy, P. S. (1972). A characterization on misspecification in the general linear regression model. *Biometrics*, *28*, 1129–1132.

Srdjevic, B., Medeiros, Y. D. P., & Faria, A. S. (2004). An objective multi-criteria evaluation of water management scenarios. *Water Resources Management*, *18*, 35–54.

Thompson, M. L. (1978a). Selection of variables in multiple regression: Part I. A review and evaluation. *International Statistics Reviews*, *46*, 1–19.

Thompson, M. L. (1978b). Selection of variables in multiple regression: Part II. Chosen procedures. *International Statistics Reviews*, *46*, 129–146.

Walls, R. E., & Weeks, D. L. (1969). A note on the variance of a predicted regression analysis. *American Statistics*, *23*, 24–26.

Yoon, K. P., & Hwang, C. L. (1995). Multiple attribute decision making: An introduction. In *Quantitative applications in the social sciences* (pp. 5–15). Los Angeles, CA: Sage Publications.

CHAPTER 10

USING SPLINE FOR DEMAND FORECASTING

Jorge Luis García A.

Department of Industrial and Manufacturing Engineering
Autonomous University of Ciudad Juarez

Erwin Adán Martínez G.

Department of Industrial and Manufacturing Engineering
Autonomous University of Ciudad Juarez

OVERVIEW

In Mexican distribution channels, milk and its subproducts are delivered in cold trucks. It is a common occurrence that demand for milk is relatively uncertain. Milk products are not sold or are returned because of the forecasting errors that arise as a result of the uncertain demand. Material handling issues that occur during the distribution of milk affect product quality negatively, which is why milk distribution trucks must have quantities for distribution that match the demand closely, with very little room for error. This chapter presents an improvement project for milk distribution in Juárez, México. Demand has a high seasonal component because of the climate variations. The forecasts were made with the adjustments of a line forecasting method, called spline. We found that with this method, we obtained a better determination coefficient than what is obtained by a linear, quadratic, or even cubic methods. Results indicate that the forecasting model based on spline fitting distributors determines reliable forecasts.

INTRODUCTION

Forecasting is the estimation of a future event based on historic information (quantitative or qualitative), gathered (Kalenatic & Blanco, 1993) for the purpose of being prepared at that point in time. In the physical distribution of products, good forecasts are needed to deliver products with a competitive level of customer service. In the present world of high economic uncertainty, business turbulence, and fierce competition and business rivalries, forecasts are becoming more important for production scheduling and sales. Forecasting is the first step of the production planning process; it is the beginning of strategic planning and the basis for the design of medium- and short-range plans. With a focus on the future, organizations that generate accurate and reliable forecasts and plans might be better prepared for competition (Adam & Ebert, 1991; Buffa & Sarin, 1995; Domínguez Machuca, García González, Ruiz Jiménez, & Álvarez Gil, 1995).

Poor forecasts of product demand result in errors in production planning. This creates an accumulation of unsold inventories and shortages that lead to penalties from customers for delivery problems and general company-wide ineffectiveness, specifically, in the management of inventories, warehousing, and pricing. Poor forecasts may also add to costs related to overproduction and lost sales. Hence, selection of a forecasting method is an important technical issue in managing operations. Equally important are any adjustments and assumptions that the forecasting techniques commonly require. Measures for improving the effectiveness of forecasts are also important for distributors to develop.

Forecasting methods can be classified by three main factors: the planning horizon, the economic environment, and the forecasting methodology. With respect to planning horizon, forecasts can be for long-range strategic business planning or be for managing day-to-day operations in a business. Forecasts are influenced heavily by both the micro and the macro economic environments. And from the perspective of forecasting methodology, forecasts can be classified as quantitative and qualitative methodologies.

Quantitative forecasting methods are mathematical and statistical methods applied to describe trends and cyclic components in production data. Quantitative forecasting methods can be classified into time series and causal forecasting methods. Time series methods are used for the study of cycles in the data. Several common time series methods are used in practice, such as weighted and moving average, exponential smoothing, and seasonal indexes. Causal forecasting methods are based on data regression, main economic indicators, and econometric methods. Their purpose is the characterization of data trends over time (Adam & Ebert, 1991; Buffa & Sarin, 1995; Chase & Aquilano, 1995; Monks, 1991; Schroeder, 1992; Tawfik & Chauvel, 1992).

Qualitative forecasting methods analyze quantitative market and economic information and are often used for validating quantitative forecasts. Commonly used qualitative methods include the Delphi panel method, informed judgment from experts, life cycle analysis, and market research methods.

Generally, an engineer would want to quantify the errors that cause a poor forecast. Forecast errors are a measure of the difference between the forecast value and the actual value. The generally accepted measures of error are the average error (AE), the mean absolute deviation (MAD), the mean square error (MSE), the percent error (PE), the mean percent error (MPE), the mean absolute percent error (MAPE), and the average deviations by period (BIAS).

Traditional Quantitative Techniques Used for Data Fitting

There are many traditional techniques used for data fitting and the development of forecast models when there is only a single independent variable. If the trend is linear and x-y plots illustrate some correlation, positive or negative, between the x and y variables, then simple linear regression adjustment can be used by ordinary least squares (OLS). However, often the data fit requires the application of higher-order polynomials, such as those of a second and third order, in order to gain a better understanding of the adjustment techniques. Linear regression technique is described in the following paragraphs.

Linear Regression Models

The classic model for OLS in linear regression is represented in equation 1.

$$y_i = \beta_0 + \beta_1 x_{i1} + \ldots \beta_p x_{ip} + \varepsilon_i \qquad i = 1, \ldots n, \qquad (1)$$

where n is the sample size, number of observations, or cases to be analyzed. The variables $x_{i1}, \ldots x_{ip}$ frequently are called observed or explanatory variables and y_i is the response variable. One of the assumptions that appear in regression analysis is that e_i (errors) has a normal distribution with average zero and unknown standard deviations. The problem in linear regression is to determine a vector of parameters unknown, which are called regression coefficients (see equation 2).

$$\beta = \begin{bmatrix} \beta_0 \\ \cdot \\ \cdot \\ \cdot \\ \beta_p \end{bmatrix} \qquad (2)$$

The regression fit is applied to a set of observations that appeared in a productive process and can be represented according to equation 3, in the matrix called *Cases*:

$$Cases = \begin{bmatrix} x_{11} & \cdot & x_{1p} & y_1 \\ \cdot & \cdot & \cdot & \cdot \\ x_{i1} & \cdot & x_{ip} & y_i \\ \cdot & \cdot & \cdot & \cdot \\ x_{n1} & \cdot & x_{np} & y_n \end{bmatrix} \quad (3)$$

Nevertheless, the true values of the regression coefficients are not known, which is the reason why it is only possible to work with estimators that have confidence levels, as is seen in equation 4:

$$\hat{\beta} = \begin{bmatrix} \hat{\beta}_0 \\ \cdot \\ \cdot \\ \cdot \\ \hat{\beta}_p \end{bmatrix} \quad (4)$$

where the $\hat{\beta}_i$ estimated values are the regression coefficients. Thus, each one of these values can be multiplied by the explanatory variables for the generation of a model, as explained with equation 5:

$$\hat{y}_i = \hat{\beta}_0 + \hat{\beta}_1 x_{i1} + \ldots \hat{\beta}_p x_{ip}, \quad (5)$$

where $\hat{y}_i$ is the predicted value for x_i. The residual ε_i of i-th observation is the difference between the given value and the predicted value for the response variable, as is seen in equation 6:

$$\varepsilon_i = y_i - \hat{y}_i \quad (6)$$

The use of linear regression and OLS goes back to the time of Gauss and Lagendre and is based on the principle of the minimization of the errors previously described in equation 6. Thus, a reduced form to express the minimization errors can be made by equation 7.

$$Minimize \sum_{i=1}^{n} \varepsilon_i^2 \quad (7)$$

According to equation 7, the basic idea is to optimize the adjustment of the line with the minimum possible residual. Like OLS, the technique is well known. The reason of its popularity is very simple to understand. In those times (approximately 1800 BCE) there were no computers and the estimation of the parameters was based on calculations by means of manipulating large matrices. It is much easier to calculate today with all the new technology we have.

The criterion of estimation established in equation 7 uses the arithmetic mean of all the observations, as much of y as x, because in those days it seemed to be one of the more reasonable parameters of location for the values and by its statistical character sufficient, since it includes all values in its estimation. When Gauss introduced the normal distribution of the errors, he demonstrated that this arithmetic mean was optimal for OLS. Since then, the assumptions proposed by Gauss for OLS have turned into standard mechanisms for the generation of many of the statistical techniques, including the linear regression.

There are many parameters that can be verified in order to validate the models generated by OLS. Likewise, residuals charts observe violations assumptions of the technique. The most used are those in which the residuals are plotted versus the independent and dependent variables. These residual plots are easy to interpret due to the visualization of trends that indicate that the model can be improved by including a coefficient of second degree or higher. In the following paragraphs we present such situations.

Second and Third Order Polynomial Fit and the Advantages and Disadvantages

If we increase the order of the equation to a third-degree polynomial, we get the following:

$$y = ax^3 + bx^2 + cx + d \qquad (8)$$

This will fit four points. A more general statement would be to say that it will fit four constraints. Each constraint can be a point, an angle, or a curvature (which is the reciprocal of the radius of an osculating circle). Angle and curvature constraints are most often added to the ends of a curve and in such cases are called end conditions. Identical end conditions are frequently used to ensure a smooth transition between polynomial curves contained within a single spline. Higher-order constraints, such as "the change in the rate of curvature," could also be added. This, for example, would be useful in the highway cloverleaf interchange design (in which left turns are handled by loop roads) to understand the forces applied to a car as it follows the shape of a cloverleaf, according to set reasonable speed limits.

Bearing this in mind, the first-degree polynomial equation could also be an exact fit for a single point and an angle while the third-degree polynomial equation could also be an exact fit for two points, an angle constraint, and a curvature constraint. Many other combinations of constraints are possible for these and for higher-order polynomial equations.

If there are more than $n + 1$ constraints (n being the degree of the polynomial), we can still run the polynomial curve through those constraints. An exact fit to all the constraints is not certain (but might happen, for example, in the case of a first-degree polynomial exactly fitting three collinear points). In general, however, some method is then needed to evaluate each approximation. The least squares method is one way to compare the deviations.

Now, one might wonder why we would ever want to get an approximate fit when we could just increase the degree of the polynomial equation and get an exact match. There are several reasons for this:

- Even if an exact match exists, it does not necessarily mean that we can find it. Depending on the algorithm used, we may encounter a divergent case where the exact fit cannot be calculated or it might take too much computer time to find the solution. Sometimes we have to accept an approximate solution.
- We may actually prefer the effect of averaging out questionable data points in a sample rather than distorting the curve for an exact fit.
- According to Runge's phenomenon, high-order polynomials can be highly oscillatory. If we run a curve through two points, A and B, we would expect the curve to run somewhat near the midpoint of A and B as well. This may not happen with high-order polynomial curves, since they may have values that are very large in positive or negative magnitude. With low-order polynomials, the curve is more likely to fall near the midpoint (it is even guaranteed to run through the midpoint on a first-degree polynomial).
- Low-order polynomials tend to be smooth and high-order polynomial curves tend to be "lumpy." To define this more precisely, the maximum number of inflection points possible in a polynomial curve is $n - 2$, where n is the order of the polynomial equation. An inflection point is a location on the curve where it switches from a positive radius to a negative radius. We can also say this is where it transitions from "holding water" to "shedding water." Note that high-order polynomials will not always be lumpy; they could also be smooth, but there is no guarantee of this, unlike with low-order polynomial curves. A 15th-degree polynomial could have, at most, 13 inflection points but could also have 12, 11, or any number down to zero.

Now that we have discussed about using a low degree for an exact fit, let us also discuss what happens if the degree of the polynomial curve is higher than needed for an exact fit. This is not acceptable for all the reasons listed previously for high-order polynomials, but it also leads to a case where there are an infinite number of solutions. For example, a first-degree polynomial (a line) constrained by only a single point instead of the usual two would give us an infinite number of solutions. This brings up the problem of how to compare and choose just one solution, which can be a problem for software programs as well as the people who work on them. For this reason, it is usually best to choose as low a degree as possible for an exact match on all constraints or perhaps an even lower degree than that if an approximate fit is acceptable.

Spline Regression

Frequently, a low-order polynomial has a poor fit and when its order is increased, it does not provide any improvement because it is possible to have a large mean square error and the coefficient of determination can decrease. Sometimes, graphs of residuals show behaviors that are hard to explain. This problem is enhanced when the data shows drastic slope changes.

This problem can be approached by considering the rank of the independent variables in various segments and fitting an equation to every segment. Spline functions can be applied in this kind of situation. This process is known as polynomial fitting by segments.

Splines are order k polynomials in which a model is fitted by segments. In this adjustment, the first $k - 1$ derivates must be given at the nodes (point-of-slope change), so that the model fitted is a continuous function. The cubic spline ($k = 3$) usually is adequate for practical purposes. A cubic spline function with h nodes, $t_1 < t_2 \ldots t_h$, with the first and second continuous derivates, can be expressed as shown in equation 9.

$$E(y) = S(x) = \sum_{j=0}^{3} \beta_{oj} x^j + \sum_{i=1}^{h} \beta_i (x - t_i)^3 \qquad (9)$$

where

$$(x - t_i) = \begin{cases} (x - t_i) & si\ x\text{-}t_i > 0 \\ 0 & si\ x\text{-}t_i \leq 0 \end{cases} \qquad (10)$$

For this kind of segment adjustment, it is assumed that the node position is approximately known and can be detected by a visual inspection in a distribution graph; if the positions are parameters to be estimated, the problem is a nonlinear regression, but if they are precisely known, equation 9 can be adjusted with ordinary least squares. In spline fitting, it is hard to decide the quantity and position of the nodes, as the order of the polynomial. Hence it is suggested that there be a minimum quantity of nodes and in each segment at least four or five points (Wold, 1974). This is important to assure data continuity and to avoid an overadjustment of the model. It is also recommended that only one inflection point be used per segment, that it remain close to the nodes, and that there is no more than one point maximum (or minimum), centered in the segment. The spline basic model can be modified easily to adjust polynomials of a distinct order in some segments and to impose continuity restrictions on the nodes.

CASE STUDY: PROBLEM DESCRIPTION

In Juarez, milk distribution is made by means of cold trucks departing from a warehouse to the points of sale. Since all the points of sale are independent supermarkets or small groceries and because there are other suppliers of milk, trucks return with an average of 34% of the original load or, in extreme cases, with unknown shortages. The returned milk quickly deteriorates after it is returned to the warehouse, mainly because of the temperature changes it suffers while in transit on stacking operations in refrigeration shells, trucks, and warehouses. One of the pertinent solutions to avoid this problem is the reduction of the errors between forecast and actual demand; the first step in the case study was to analyze the forecast system being used. We observed two problems in the actual system. First, in the company, previous years' data were discarded; hence historical data on any annual trends in milk distribution were not available. Forecasts were moving averages, which were calculations of 3 months from the previous year. For instance, the forecast of the March 2009 demand was calculated with demand data from the first 3 months of 2008. Therefore, seasonal behavior was not observed with the calculation of the moving averages. We also considered that planning in monthly periods was ineffective.

METHODOLOGY

The first task was the construction of the dispersion diagram, illustrated in Figure 10.1, where the independent variable was the previous year's sales divided in 52 weeks. As can be observed, there are two slope changes. Given that the information available was from just 1 year, it was not possible to develop the common regression equation for annual trend analysis.

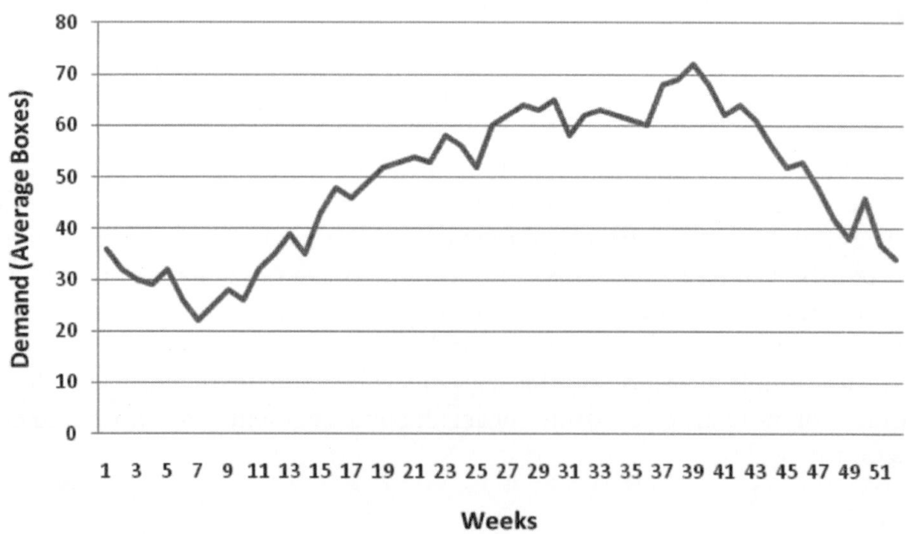

Figure 10.1. Dispersion diagram of weekly sales.

We adjusted the square and cubical type, and it was observed that the obtained mean squared for residuals from an adjusted cubical spline was smaller than the other polynomial fits, which consist of fitting line segments with the unique condition that the nodes must continue. So, this procedure was applied.

The software used to solve and generate the models was NCSS. Minitab was used to generate the graphs. The spline was generated using transformations for independent variables in Minitab.

The only indicators of efficiency for the generated models were the determination coefficient (R^2) and the square sum of residuals (SSE). Values close to one of the determination coefficients are desired, and low values for SSE are required.

APPLICATION EXAMPLE

Table 10.1 shows last year data of the weekly sales of milk. The dispersion diagram shown in Figure 10.1 presents two inflection points and two slope changes that we visually estimate in points 7, 22, 39, and 79. Four adjustments were made for the comparison of the values of the multiple determination coefficient (R^2) and the mean square error: linear, quadratic, cubic, and spline. Estimates are illustrated in following paragraphs.

Linear Model

The linear model adjusted to this data and its parameters appears in Table 10.2. A residual graph versus the dependent variable appears in Figures 10.2a and 10.2b. According to those parameters, the linear model is inefficient.

Second-Order Model

The second-order model adjusted to the data and its parameters appears in Table 10.3. A residual of demand graph versus the predicted demand in Figure 10.3 is also shown. A graph that illustrates dependence among data appears in Figure 10.4. It is easy to see that the model requires some improvement in some parameters, such as in the MSE and in the R^2. Also, according to those parameters, the linear model is inefficient due to trend in residual graphs.

Table 10.1. Week Sales

Week	1	2	3	4	5	6	7	8	9	10	11	12	13	14	15	16	17	18
Sales	36	32	30	29	32	26	22	25	28	26	32	35	39	35	43	48	46	49
Week	19	20	21	22	23	24	25	26	27	28	29	30	31	32	33	34	35	36
Sales	52	53	54	53	58	56	52	60	62	64	63	65	58	62	63	62	61	60
Week	37	38	39	40	41	42	43	44	45	46	47	48	49	50	51	52		
Sales	68	69	72	68	62	64	61	56	52	53	48	42	38	46	37	34		

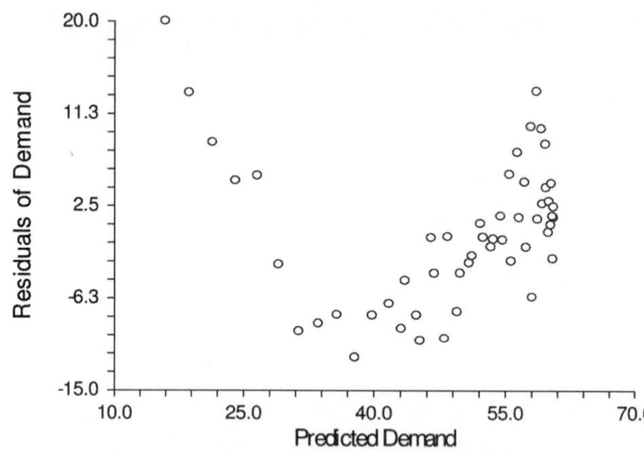

Figure 10.2a. Residuals of demand versus predicted demand for simple linear regression model.

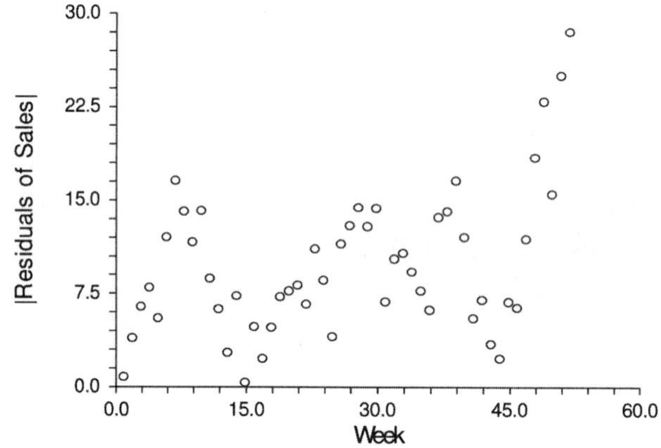

Figure 10.2b. Residuals of sales versus week.

Table 10.2. Parameters for Linear Model Adjusted

Parameter	*Value*
Intercept	34.7466
Slope	0.5328
R-squared	0.3324
Correlation	0.5766
Mean square error	133.5444

Table 10.3. Parameters for Second-Order Model

Parameter	*Value*
Intercept	13.1305
Week	2.99346
Week^2	–0.04553
R-squared	0.7653
Mean square error	47.98

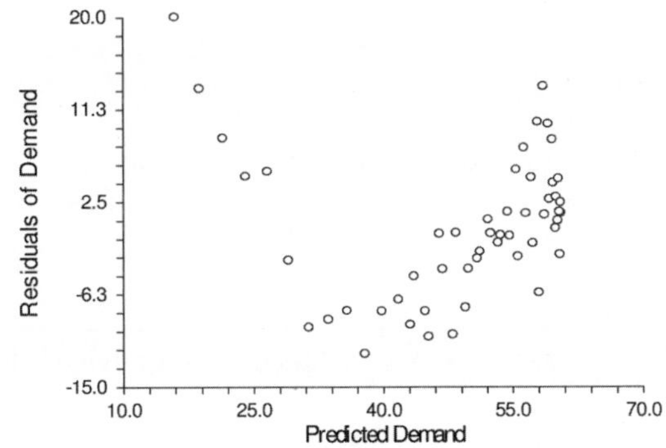

Figure 10.3. Residuals of demand versus predicted.

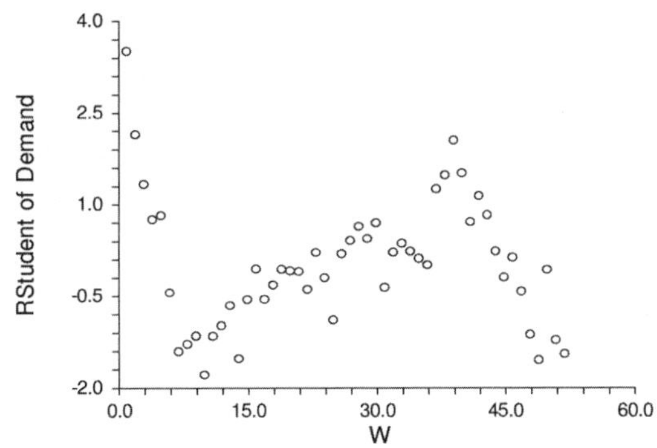

Figure 10.4. R student versus week.

Third-Order Model

The third-order model adjusted to data and its parameters appears in Table 10.4. A residual graph versus the independent variables and predicted values for the residual analysis made is also presented. It is easy to see that the model has some improvement in some parameters, like MSE and R^2, because the MSE was reduced and R^2 was increased, which means that the model has an error and is explaining more variance. According to those graphs and parameters, we can conclude that the model is more efficient than the others of the first and second order but may be improved using cubic spline.

According to Figure 10.5, the residual does not show any trend and looks homogeneously distributed along the predicted values, which means that the model can be accepted.

Cubic Spline

The spline model adjusted to data and its parameters appears in Table 10.5. A residual graph versus the independent variables and predicted values is also shown. It is easy to see that the

Table 10.4. Parameters for Third-Order Model

Parameter	*Value*
Intercept	29.1842
Week	–0.5360
$Week^2$	0.1168
$Week^3$	–0.0020
R-squared	0.9172
Mean square error	17.24

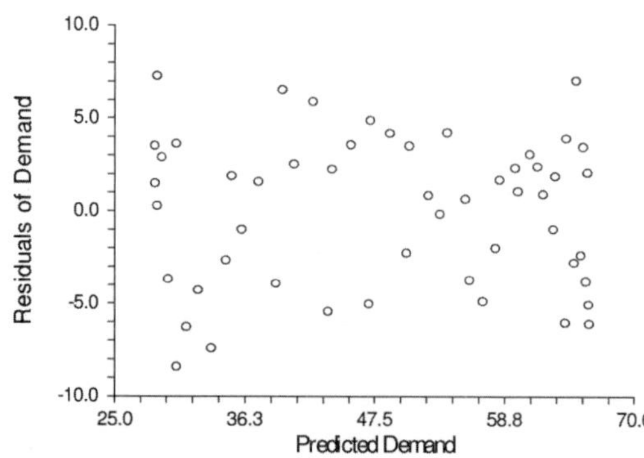

Figure 10.5. Residual versus predicted.

Table 10.5. Spline Model Parameters

Parameter	*Value*
Intercept	46.544
(Week)	–13.799
$(\text{Week})^2$	3.793
$(\text{Week})^3$	–.33898
(Week 7)	14.353
$(\text{Week 7})^2$	3.1901
$(\text{Week 7})^3$	0.3406
(Week 39)	–2.2795
$(\text{Week 39})^2$	–0.21402
$(\text{Week 39})^3$	0.0110
R-squared	0.9653
Mean square error	8.2864

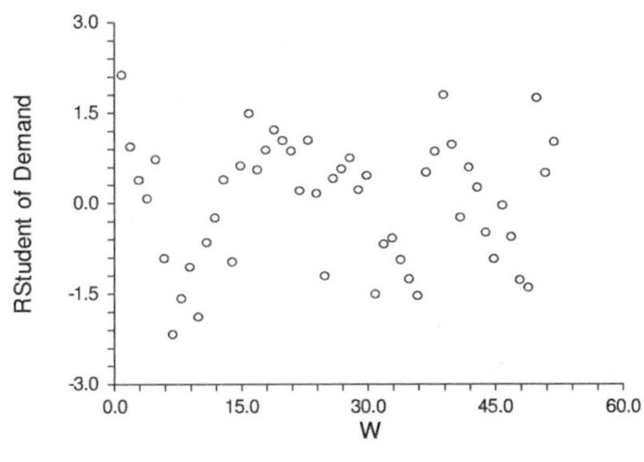

Figure 10.6. R student versus week for third-order model.

model has some improvement in some parameters, particularly the MSE and the R^2. According to the graphs and parameters, the model is efficient and produces less error in forecast.

Finally, in Figure 10.9, the trends that have the efficiency index of the coefficient of determination appears. From this figure, it is easy to observe the accelerated growth that has this value when it shifts from one model to another. In this case, the linear model has the worst indicator and the spline-generated model has the best. Similarly, Figure 10.10 shows graphically the decline growth that has the MSE indicator. In this case, it appears that the value obtained by the linear model is the worst of all and the spline model is the best fit.

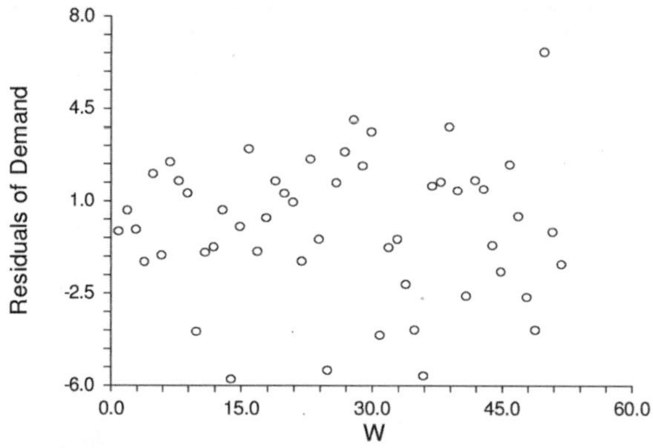

Figure 10.7. Residuals versus week.

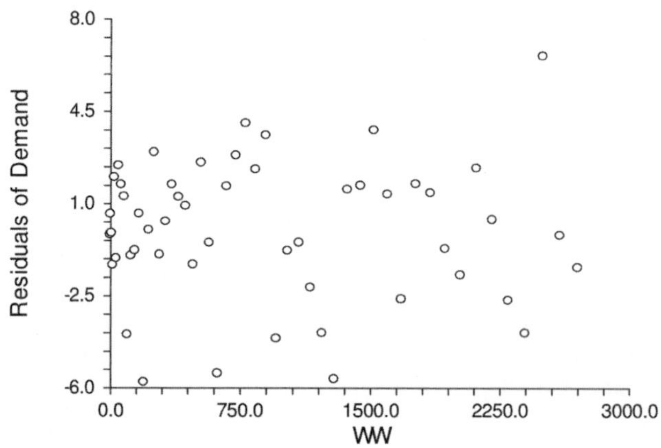

Figure 10.8. Residuals versus week^2.

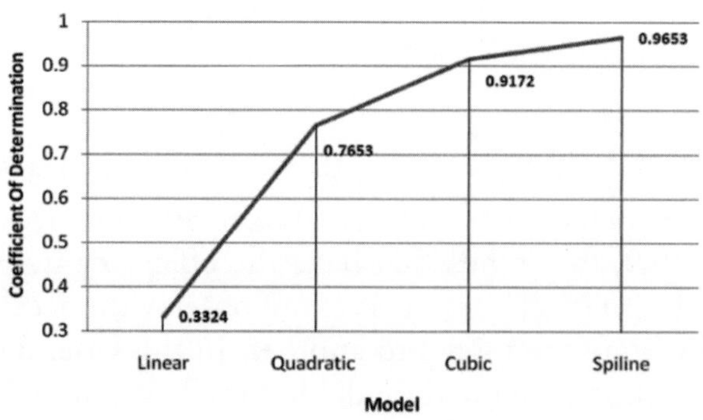

Figure 10.9. Models and their R^2.

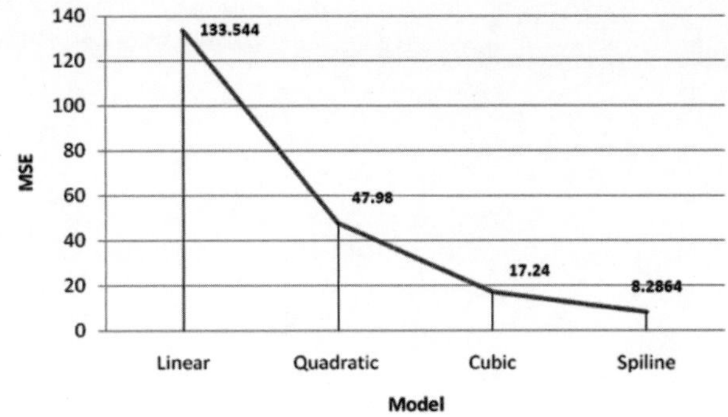

Figure 10.10. Models and their MSE.

CONCLUSION

Given the results of this project, we have shown that spline adjustments can improve the effectiveness of a forecasting model based on OLS regression. The application of this method is relatively simple, so small companies may be interested in its use. Calculations and adjustments can be made in Excel worksheets and software such as NCSS. It is desirable for further tests of this method to be performed to obtain empirical evidence of its effectiveness.

REFERENCES

Adam, E., & Ebert, R. (1991). *Production and operations management* (4th ed.). Mexico City, Mexico: Prentice Hall.

Buffa, E., & Sarin, R. (1995). *Production and operation management*. Mexico City, Mexico: Limusa.

Chase, R., & Aquilano, N. (1995). *Production and operations management* (6th ed.). Barcelona, Spain: IRWIN.

Domínguez Machuca, J. A., García González, S., Ruiz Jiménez, A., & Álvarez Gil, M. J. (1995). *Operations mangement: Tacticals and operatives in services*. Madrid, Spain: McGraw-Hill.

Kalenatic, D., & Blanco, L. E. (1993). *Computational applications in production*. Fondo Santa Fé de Bogotá, Colombia: Universidad Distrital Francisco José de Caldas.

Monks, J. (1991). *Operations management*. Mexico City, Mexico: McGraw-Hill.

Schroeder, R. (1992). *Operations management, decision-making in the operations function* (3rd ed.). Mexico City, Mexico: McGraw-Hill.

Tawfik, L., & Chauvel, A. M. (1992). *Production management*. Mexico City, Mexico: McGraw-Hill.

REFERENCES

CHAPTER 11

MAINTENANCE DATA MANAGEMENT: ACQUIRING AND CREATING KNOWLEDGE FOR MAINTENANCE ACTIVITIES

Masaya Hagiwara

Department of Mechanical Engineering
Nagoya Institute of Technology

OVERVIEW

In this chapter, the database system for maintenance management information on the Internet is described. This system will offer information including fault identification, selection of adequate maintenance policies, and evaluation of the maintenance activities. The system proposed is based on the hierarchical fault classification (HFC) method, by which an undesirable event, in other words, a failure that occurs in the equipment in manufacturing plants, is expressed by using a series of fault modes of eight hierarchical levels from the generic cause level to the society level. Any fault modes in any hierarchical levels can be identified in the single formulated manner by using this information, including fault level, item name, faulty state, and the method of the fault detection or identification connected. Consequently, a specific failure that occurred in a manufacturing plant can simply be described by connecting the upper and lower fault modes together. By using such a concept, the plant personnel in various areas can share the necessary technical data to be used for their own maintenance management or activities. A small-scale database system that will be presented has proved the possibility to realize the multiplicative database based on the HFC method that could offer the necessary information via the Internet.

INTRODUCTION

Maintenance plays an important role in highly automated manufacturing plants to save on resources including energy resources, human resources, and material resources and to maximize the productivity from an engineering point of view. Maintenance is the combination of all technical and administrative actions to reduce the consequences of failures, extend the life of the system, and ensure that the condition of the system meets all authorized requirements (Kumar, Crocker, Knezevic, & El-Haram, 2000). Therefore, the efficiency of maintenance activities is extremely important. It can prove difficult to perform maintenance effectively and efficiently since it requires a great number of concrete and technical data or knowledge concerning fault diagnosis in a broad sense (e.g., fault identification, failure assessment [cause and effects analysis], and fault correction or prevention). This means that maintenance data management is one of the most important keys to make the maintenance activity effective and efficient.

Many studies and efforts have been performed to formulate the effective maintenance management system. For maintenance project control, program evaluation and review technique (PERT) and critical path method (CPM) are widely used (see Dhillon, 2002), and the concept of total productive maintenance (TPM) is combined with reliability centered maintenance (RCM) to save the resources in manufacturing (Smith & Hawkins, 2004). Furthermore, an intelligent maintenance system called Watchdog Agent®-based real-time remote machinery prognostics and health management (R^2M-PHM) system has been recently developed (Kobbacy & Murthy, 2008). However, these tools do not always work successfully in some types of practical manufacturing plants in terms of effectiveness and efficiency in maintenance, since the performance of maintenance is highly dependent upon the types and the characteristics of equipment, machinery, and systems (Desai & Mital, 2005). The plant personnel sometimes have to get the data described above for their own items by themselves. In fact, a great number of valuable technical data have been produced and published through investigations or research works done by the experts from various areas. However, these data cannot be fully utilized in the maintenance management area since most of the data are not integrated as useful knowledge for maintenance management but are only scattered as the elements or the row materials described in different formats.

In this chapter, the database system is described for maintenance management including fault identification, selection of adequate maintenance policies, and evaluation of maintenance activities, which allows the plant personnel from various types of manufacturing plants to share the basic and highly specialized data. For this purpose, a new concept is proposed that will offer the necessary technical information and the knowledge for individual plant personnel via the Internet.

The database system is based on the hierarchical fault classification (HFC) method (Chen & Hagiwara, 2000; Hagiwara, 1996) by which all of the data can be shared and utilized by both a large number of experts and users in various areas. The data module formulated to express the general fault diagnosis information is proposed, and the functionalities of this method are highlighted using some examples.

MAINTENANCE DATA AND THEIR DESCRIPTIONS

Classifications on the Data to Be Used for Maintenance

Figure 11.1 shows the key issues in the maintenance of an asset (Kobbacy & Murthy, 2008). To execute the maintenance action, one should understand and utilize the concepts and techniques and therefore needs a variety of data or knowledge for each of the issues.

Technical data (maintenance data) are required at the faulty state recognition stage and the planning and execution of the maintenance activity stage (Hagiwara, 1999). The following explains these two stages:

1. Data for failure or faulty state recognition
   a. Faulty item (item names and the locations)
   b. Faulty state (type and other features)
   c. Fault recognition or monitoring technique (methods, equipment or tools, and procedures)
   d. Fault detection or diagnosis technique (method, equipment or tools, and procedures)

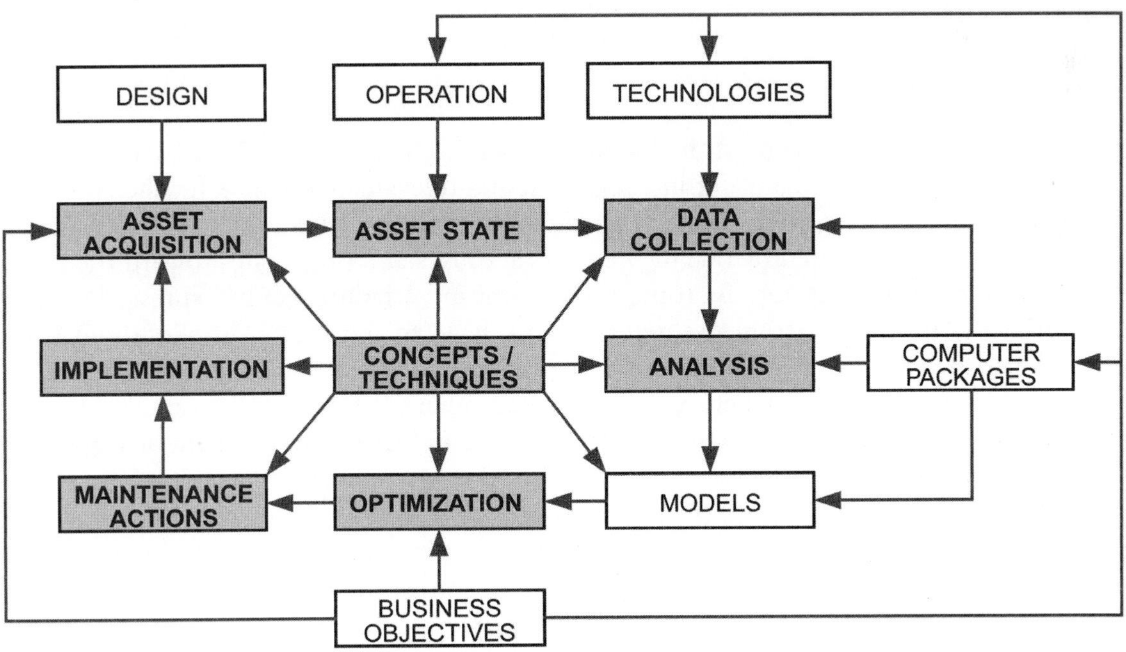

Figure 11.1. Key issues in maintenance of an asset.
Source: Kobbacy and Murthy (2008).

2. Data for the maintenance planning and execution

    a. Effect of the failure (criticality and damage quantified)
    b. Type of the failure (period, frequency, and other features)
    c. Access technique to the faulty item(s) (methods, tools, and procedures)
    d. Restoration or correction technique to the faulty item(s) (methods, tools, and procedures)

These data may contain some inherent or local issues for the individual plant or equipment. However, they are very restrictive and only affect significantly the data presented in the 2a and 2c lines, concerning the effect of the failure and the access technique or method to the faulty item in an individual plant. Therefore, we can build up the maintenance data utilized for our own maintenance activities from published knowledge or the elements of the technical data. The most important thing to do now is to exclude the difficulties in building up the data set due to the wide varieties of the descriptions of knowledge.

Failure modes and effects analysis (FMEA) and fault tree analysis (FTA) are typically used to assess the effects of a failure and what the failure (root) causes (e.g., Levin & Kalal, 2003). Root cause analysis is a newly developed technique, and Ray, Elam, and Bledsoe (2007) have concretely applied it to rocket engine failures to boost the reliability. However, these techniques are designed to be applied to a specific or individual system concretely rather than to formulate or standardize the description of the technical data for the maintenance management in general.

Concept of HFC

Figure 11.2 shows the concept of the hierarchical fault classification (HFC) system used to describe the faulty states that occur in equipment and to evaluate the maintenance policy to be applied (Chen & Hagiwara, 2000; Hagiwara, 1996). The HFC method is based on the fact that a failure is initiated by one of the generic causes and propagates toward the upper hierarchical levels in each step. By using such a concept, a failure can be expressed by a set or a series of fault modes from the generic cause level to the society level without using fuzzy words such as "cause" and "effect." The plant is divided into five hierarchical levels in accordance with the system structure: plant (P), equipment (E), unit (U), assembly (As), and component (Cp) levels. Furthermore, society (S) level and stress (St) and generic cause (Gc) levels are defined as upper and lower levels of the system structure, respectively.

By using the HFC concept, the maintenance activity can be expressed in a simplified manner. When performing the maintenance activity, you must "access the item of lower level step by step until the failed item to be replaced has been reached, and replace it."

The equipment can be restored by replacing one of the upper-level items (assembly, unit, or equipment) instead of replacing the failed component or the failed part of the component. In general, if you select the upper maintenance level, the damages and the maintenance item cost increase, but the maintenance work would be easier. If you contrarily choose

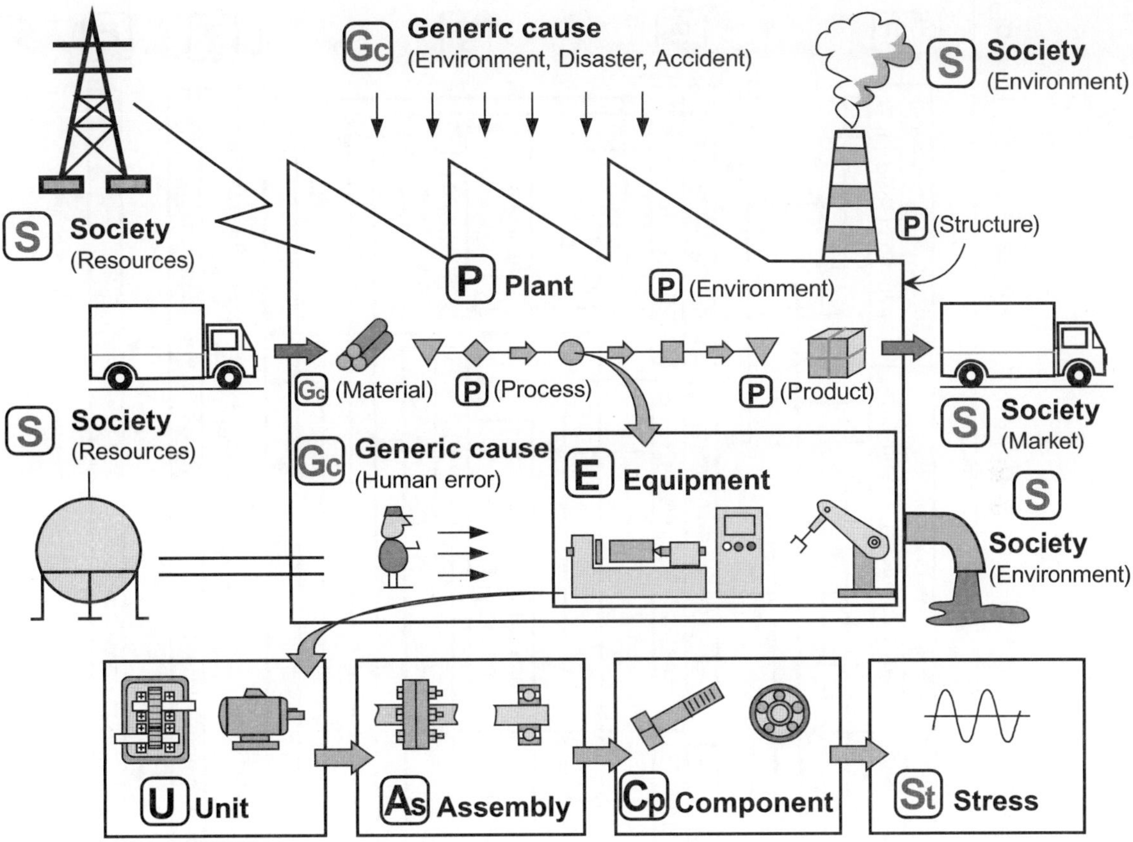

Figure 11.2. The concept of *level* and *item* in the HFC.

the lower maintenance level including the exclusion of the root cause (fault mode of the generic cause level), the damages and the maintenance item cost would be minimized, but the technical difficulty increases, and it may result in higher total cost.

Expression of Maintenance Data by Using HFC Method

Some features and advantages of HFC are highlighted by two examples:

Example 1: Expression of the Route Map *Showing All the Possible Failures in a System*

Figure 11.3 shows an example of the description of the possible failures that may occur in bolted or threaded joints in the general machinery. This diagram lists all of the failure mechanisms or the relationships between the causes and the effects that could also be obtained by

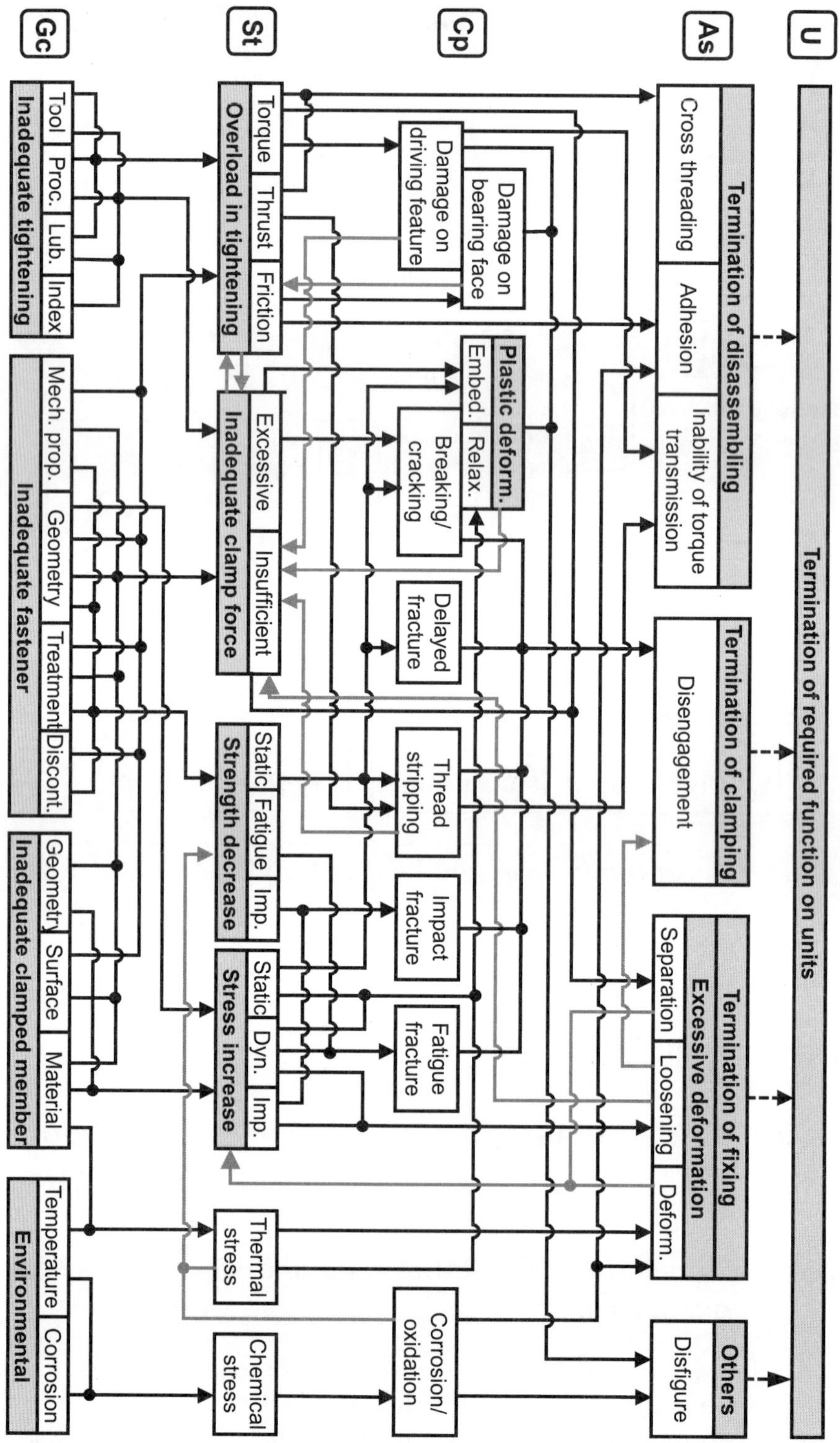

Figure 11.3. An example of the expression of possible failures in bolted and threaded joints.

using the conventional techniques such as FTA and FMEA. In HFC, however, each failure can be described objectively by using the hierarchical levels as "absolute coordinates." Items and their faulty states can be classified easily into one adequate hierarchical level in advance. The primary failures always start from one or more generic cause level fault mode(s) and spread toward the upper-level fault modes. In some cases, a primary failure generates an excessive stress (denoted as the stress level fault mode) for other component(s), and the secondary failure occurs then.

By collecting the same kind of information as in Figure 11.3 for other types of components and assemblies, we can list all the possible unit level fault modes, and they could be connected to the equipment, the plant, and the society level fault modes.

Example 2: Expression of a Specific Failure With Fault Diagnosis (Assessment) Information

Figure 11.4 shows an example of the failure assessment performed on a specific event. The operator was first aware of the noise and the chattering in a cylindrical grinding machine, and he shut down the machine. The assessment was started from two fault modes of the equipment level to both upper and lower levels. Consequently, it was clarified that the root cause (the generic cause level fault mode) is the inadequate selection of the lubricant for the main spindle bearing, and the fatigue fracture (secondary failure) occurred on the race of that bearing. Then the most adequate method for detecting or monitoring each fault mode was considered, and finally, it was decided that the composition of the lubricant would be changed and that the condition monitoring method using FFT would be introduced.

Thus, in HFC, a path on the diagram represents the failure propagation, and the adequate maintenance policy can be evaluated and chosen by answering the following single question: At which level should we cut the path(s) for the failure to be considered?

CONCEPTUAL DESIGN FOR THE DATABASE SYSTEM

Contents and Form of Data

As mentioned earlier, the maintenance data or information utilized for the maintenance management for a certain failure can be summarized in the form of the diagram such as shown in Figure 11.4. However, not all plant personnel have sufficient knowledge to make such a diagram. Therefore, the database should be designed to offer the necessary data or the data elements to complete the diagram. The failure roadmap in Figure 11.3 is useful information to make the diagram in Figure 11.4.

The information shown in Figures 11.3 and 11.4 consists of data modules that have a couple of fault modes and their relations. Figure 11.5 shows the general form of a data module. A primary failure and a secondary failure start from the generic cause level fault mode and the stress level fault mode, respectively, and propagate to the upper level. This means

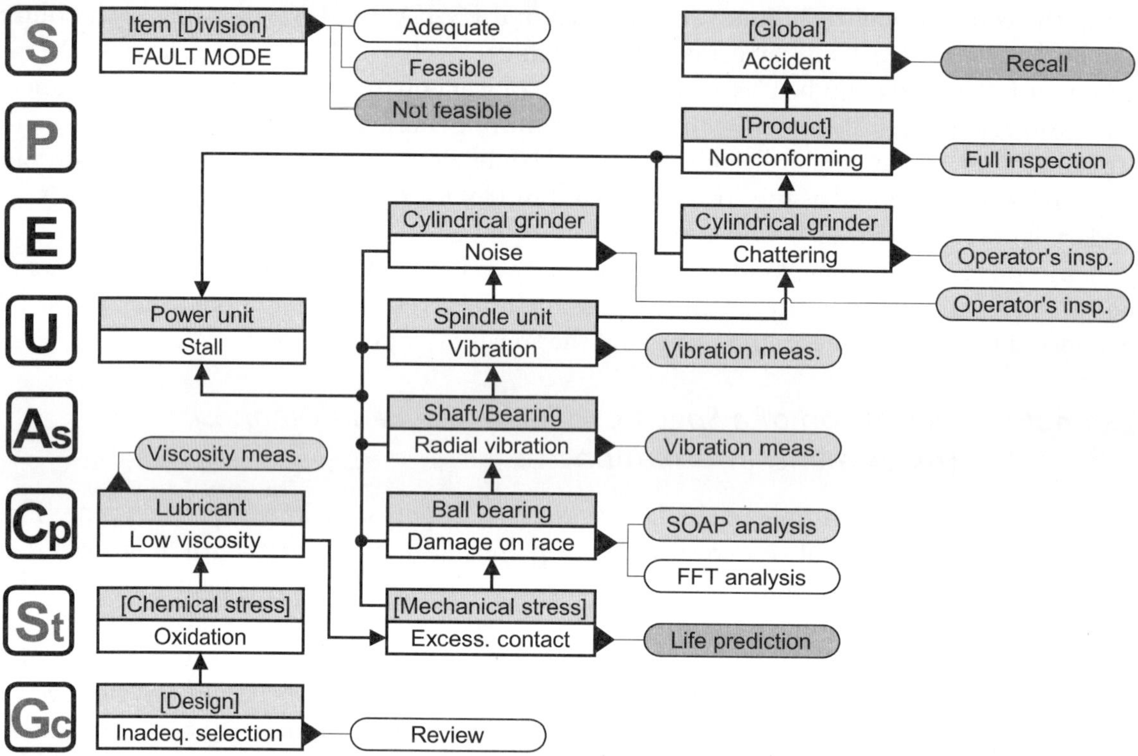

Figure 11.4. An example of the expression of a specific failure with its assessment in a cylindrical grinding machine.

that a failure path can be obtained by "pasting" the appropriate data module upward and downward until the society level fault mode, the stress level fault mode (cause of secondary failure), and the generic cause level fault mode have been reached.

Database Structure

Figure 11.6 shows an example of the database structure that enables one to store the information in the form of the data module as shown in Figure 11.5. To avoid the overlapping and double definition of the data, item and fault mode names should be standardized by predetermined terms registered in the library. For this purpose, the detailed classification code system such as in ISO 13584–511 (2006) is not always necessary, and for the time being, the simpler one is thought to suffice (e.g., see Figures 11.7a and 11.7b).

Concerning the fault mode or faulty state name, the simple classification criteria such as used in FMEA (IEC 812, 1985) could also be applicable since the data module itself represents the failure mechanism information, while the component fault mode name conventionally used contains the failure mechanism such as "*fatigue* fracture" and "*stress corrosion*

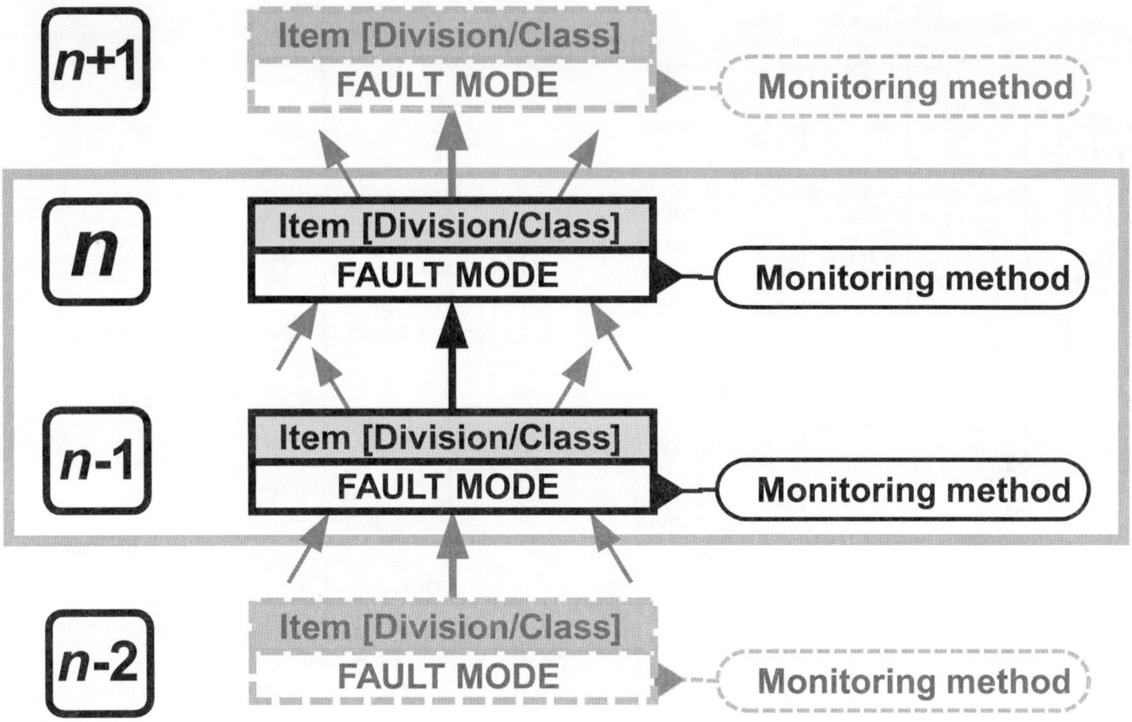

Figure 11.5. Structure of data module.

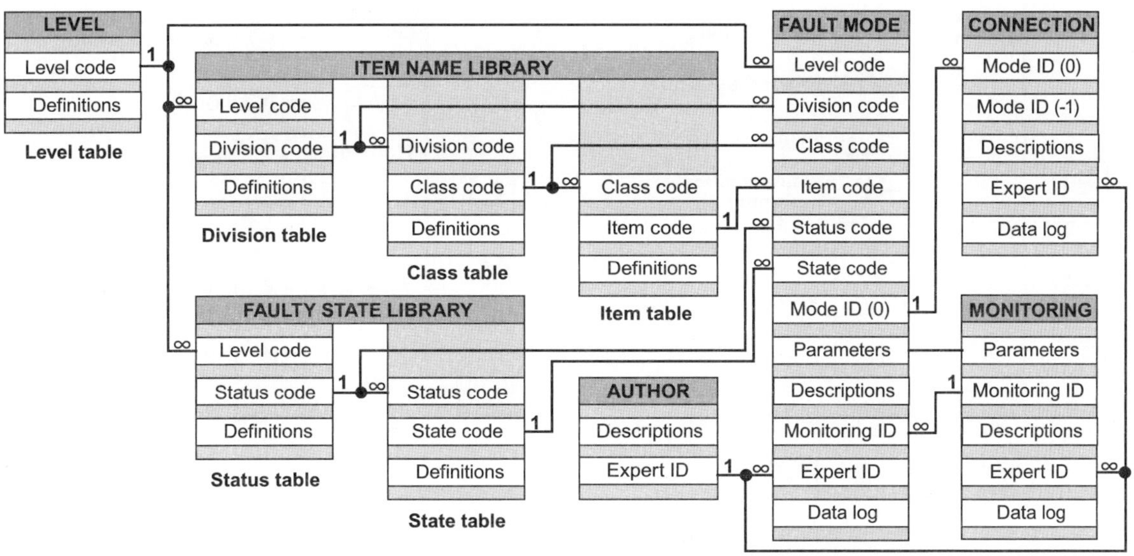

Figure 11.6. Basic structure and relationships of the database system.

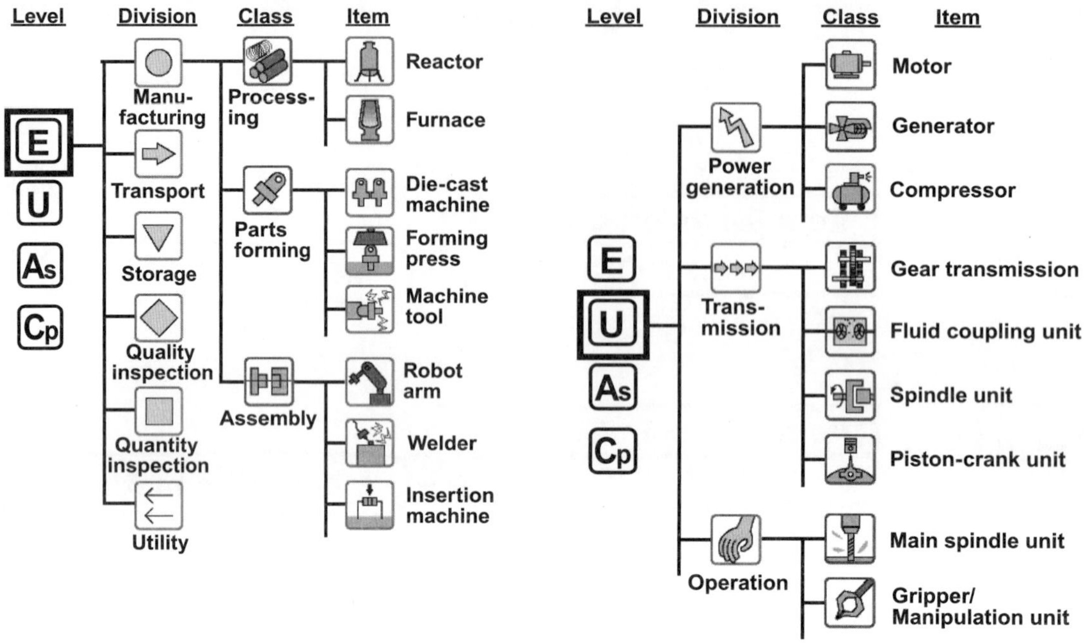

Figure 11.7a. An example of the criteria for classifying items (equipment and unit levels).

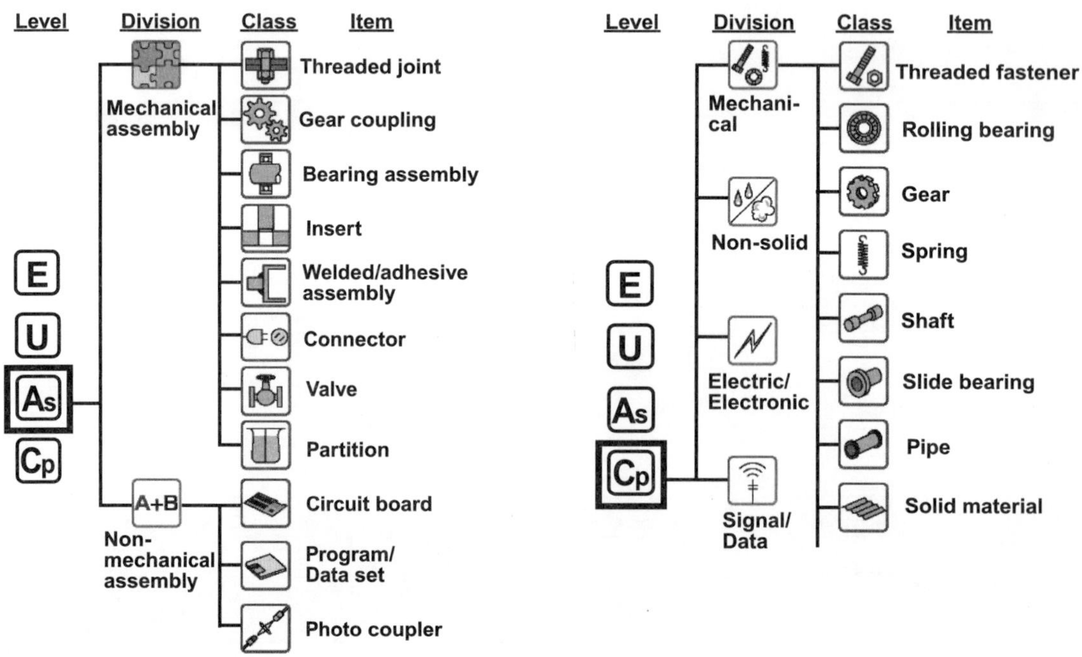

Figure 11.7b. An example of the criteria for classifying items (assembly and component levels).

cracking." However, a certain kind of redundancy might be helpful for the practical system to be human friendly.

Method of Data Collection and Operation

Table 11.1 shows the classifications of the typical experts who will offer the data. The experts will present data concerning the technical terms on the item and fault mode, the data modules, or the monitoring techniques on their area of expertise. For performing such editorial works efficiently, the authors should refer to the data that have already been stored.

On the other hand, the users would refer to and gain the data to make their own information such as what is shown in Figure 11.4. If the statistics (frequencies) to choose the individual data module would be counted and collected automatically, we could offer the possibilities of certain types of failures (the data on line 2b earlier) as support information.

As for the presentation of the data, there are many possible ways. An example is shown in Figure 11.8 where the fault mode of the disengagement of a bolted joint (As level) is explained, and the bolt breaking (Cp level), the shear stress on thread surfaces (St level), and the stripping of the nut thread (Cp level) are listed as the possible direct causes for the disengagement of a bolted joint (As level).

Table 11.1. Classifications of the Areas and the Experts Offering the Technical Data

Fault level	Item	Major areas	Experts
S Society	Human	Medicine Medical eng.	Researcher (University, Institute) Design/manuf. engineer Maintenance engineer Condition monitoring engineer
	Creature/nature	Environmental eng. Agriculture	
	Social assets	Architecture Civil eng.	
P Plant	Manufacturing systems	Systems eng. Industrial eng. Management eng. Design eng.	
E Equipment	Equipment		
U Unit	Functional units	Information eng. Mechanical eng. Material eng. Electronic eng. Chemical eng.	
As Assembly	A couple of components		
Cp Component	Parts/components		
St Stress	Microstructure	Material science Physics Chemistry	
Gc Generic cause	Plant personnel	Ergonomics	
	Infra-structure	Architecture Civil eng.	
	Environment	Environmental eng.	

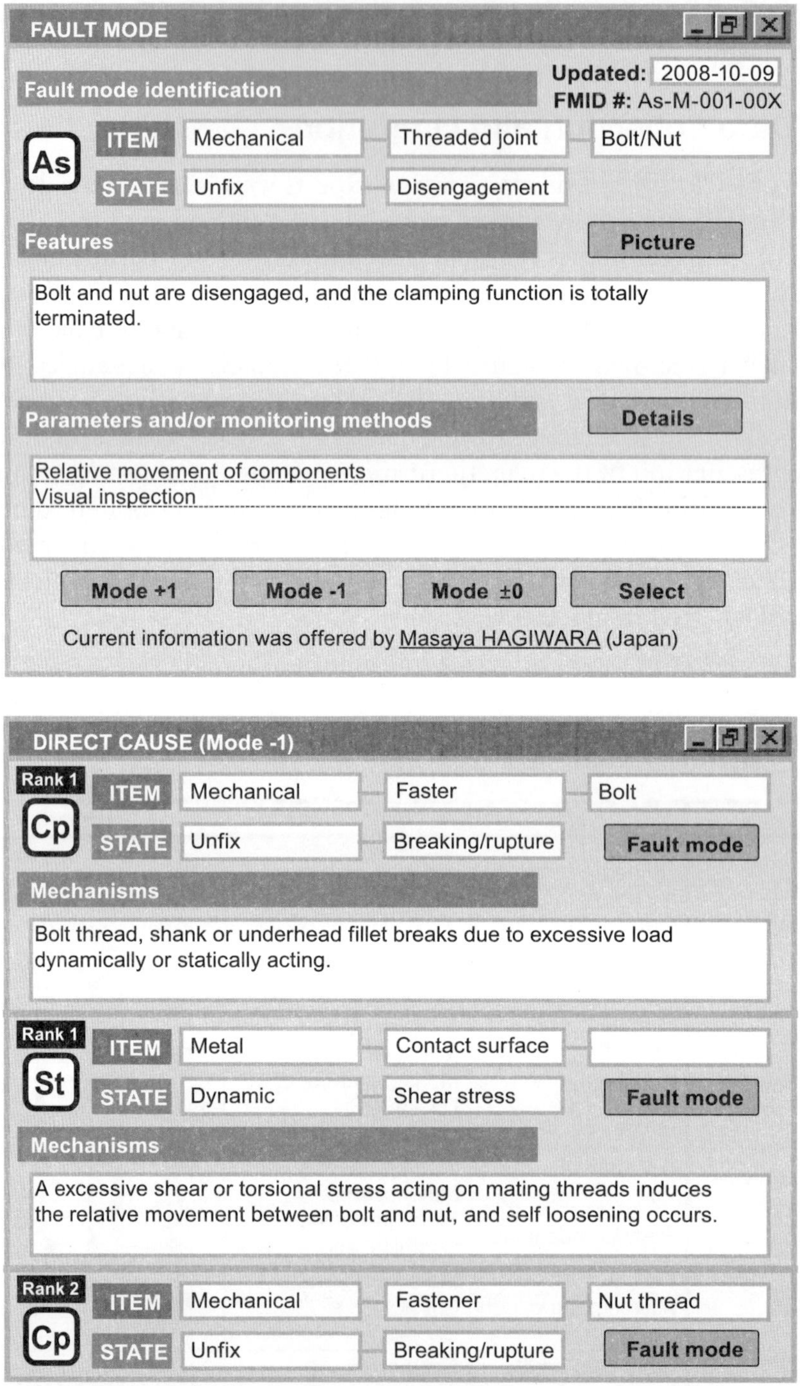

Figure 11.8. An example of the data format showing a fault mode and the possible direct cause.

CONCLUSION

In this chapter, a concept of the database system was proposed by which plant personnel can gain the necessary technical data for maintenance management in their manufacturing plants, including information concerning failure assessments, choices available for an adequate maintenance level, and actions to be performed.

It was emphasized that the HFC method is the most appropriate methodology to standardize the maintenance information and to make a human-friendly database for both experts and users to share the knowledge. Some examples, including the structure of the database and operational issues, were presented to highlight the advantages of the system proposed.

REFERENCES

Chen, J., & Hagiwara, M. (2000). Development of a maintenance management support system based on a hierarchical fault classification. *International Journal of Industrial Engineering, Theory, Applications and Practice*, 7(3), 255–264.

Desai, A., & Mital, A. (2005). State of the research on design for maintenance algorithms. In *Proceedings of the 10th Annual International Conference on Industrial Engineering—Theory, Applications and Practice* (pp. 245–250). Cincinnati, OH: IJIE.

Dhillon, B. S. (2002). *Engineering maintenance: A modern approach*. Boca Raton, FL: CRC.

Hagiwara, M. (1996). A new technique in maintenance management based on hierarchical fault classification. In J. J.-G. Chen & A. Mital (Eds.), *Advances in industrial engineering applications and practice I* (pp. 1060–1065). Cincinnati, OH: IJIE.

Hagiwara, M. (1999). Data sharing technique for maintenance management. *Journal of the Japan Society for Precision Engineering*, *65*(3), 377–380.

ISO 13584-511. (2006). In *Industrial automation systems and integration—parts library—part 511: Mechanical systems and components for general use—reference dictionary for fasteners*. Geneva: International Organization for Standardization.

Kobbacy, K. A. H., & Murthy, D. N. P. (Eds.). (2008). *Complex system maintenance handbook*. London, England: Springer Verlag.

Kumar, U. D., Crocker, J., Knezevic, J., & El-Haram, M. (2000). *Reliability, Maintenance and Logistic Support—A Life Cycle Approach*. Norwell, MA: Kluwer Academic Publishers.

Levin, M. A., & Kalal, T. T. (2003). *Improving product reliability: Strategies and implementation*. Chichester, England: John Wiley & Sons.

Ray, P. S., Elam, M. E., & Bledsoe, A. M. (2007). Root cause analysis—beyond failure mode and effects analysis. *In Proceedings of the 12th Annual International Conference on Industrial Engineering—Theory, Applications and Practice* (pp. 471–475). Cincinnati, OH: IJIE

Smith, R., & Hawkins, B. (2004). *Lean maintenance: Reduce costs, improve quality, and increase market share*. Oxford, England: Elsevier Butterworth-Heinemann.

CHAPTER 12

SHELF MANAGEMENT IN APPAREL INDUSTRY

Jung-Wook Lee

Department of Management Science
Korea Advanced Institute of Science and Technology

Yoon-Min Hwang

Department of Management Science
Korea Advanced Institute of Science and Technology

Jae-Jeung Rho

Department of Management Science
Korea Advanced Institute of Science and Technology

OVERVIEW

Currently, radio frequency identification (RFID) is considered the most promising technology in apparel retailing. It not only enhances the retail operation but also improves customer satisfaction. Focused on the customer shopping process, this study examines how RFID data is applied to shelf management to determine optimal product assortment and shelf space allocation to meet the unpredictable customer demand. To do this, the study first reviewed literature on the shopping process and used important findings of shelf management in an online environment to explore the use of RFID data as an offline log data. Next, we designed the systems and process to execute RFID-based shelf management in apparel retailing. Finally, the study suggests guidelines for shelf management decision making using RFID data.

INTRODUCTION

In the retail business, shelf management is a decision-making process designed to maximize profit by arranging various products optimally under the limited space before sales start (Borin & Farris, 1995). In the case of apparel retailing, so much importance has been attached to shelf management because of certain characteristics of fashion products—high demand volatility, short sales period, and high variety of products (Rajaram, 2001). According to a review recently conducted in an Australian department store, footwear was capable of delivering the desired sales using only 65% of its allocation space, while 37% of men's wear in that same store was not properly displayed or easily accessible to the customer. It also found that approximately 22 completely different products are presented to the customer on three display hooks (CapturePlan, 2005). This means that retailers failed to sort and to display some products for sale.

Currently, retailers determine how and where retail products should be displayed using a tool called planogram. It is a collection of detailed maps that show store personnel exactly where all the products are to be displayed in the store considering shelf capacity, product type, and consumer point of sales (POS) information (Rao, 2000). However, nothwitstanding the use of state-of-the-art technology, shelf management in apparel retailing is still struggling due to the shortage of information about new products and customers' attitudes when the season changes (Ghani, 2004). For this reason, few studies suggest alternatives such as a surveillance camera or direct observation to grasp customer preference (Newman & Cullen, 2002). However, these kinds of solutions have some limitations in terms of cost, technology, and ethics (Ahn, 2005).

Recently, with the advancement and deployment of RFID technology and the desire for quick response to customer demand, apparel retailing utilizes the attachment of sensors to items, which provide real-time monitoring of merchandise on shelves as well as in the back rooms of the store (Kim & Kim, 2006). Collected RFID data would be applicable not only to retail operations but also to new businesses with the opportunity to grasp customer preference as offline log data (Ahn, 2005). Since the key in the apparel market is to monitor and react to customer demand, the Metro Group has tested the technology on their store by offering a novel experience through smart dressing rooms. These dressing rooms provide product information on individual items with regard to what customers wish to try on (Wessel, 2007).

This study examines how RFID data is applied to shelf management to meet the unpredictable customer demand. To do this, first we reviewed literature on the shopping process and used important findings of shelf management in an online environment to explore the use of RFID data as offline log data. Second, the systems and process to execute RFID-based shelf management are designed for apparel retailing. This study suggests guidelines for decision making using this RFID data.

LITERATURE REVIEW

Shelf Management

Shelf management consists of two steps: product assortment and shelf space allocation. When new products are released, retailers have to determine the number and types of products in a line (Rajaram, 2001). In other words, the line carries the number of stock keeping units (SKUs) to each product category. Then products are assigned to space on shelves considering their width, relationships of complementary and substitute items, and the shelf level.

A related problem has been developed considering some verified effects on purchasing in a retail environment. For shelf space allocation, main space elasticity—the ratio of relative change in unit sales to the shelf space—is proposed first (Anderson & Amato, 1974). Also location elasticity reflects the level of shelf where the product is displayed (Drèze, Hoch, & Purk, 1994), while cross-space elasticity, which considers not only shelf space management but also product assortment, deals with complementary and substitute relationships between products (Corstjens & Doyle, 1983). Corstjens and Doyle and Urban (1998) studied this problem, including the inventory situation.

However, related studies focus on the use of historical data for shelf management prior to the season. Therefore, they do not highlight the importance of shelf management during the season or between two seasons.

RFID Technology

RFID is an identification technology that uses radio signals to read data stored in a tag, also called a smart chip, at a remote location with no direct line of sight and without labor. Compared with barcode technology, RFID can scan a product remotely and automatically in bulk; therefore it performs faster and retains more identification per unit time than a barcode. Once a tag is attached to an object, such as a pallet, box, or item, each individual logistic unit would be identified and seamlessly traced by many readers installed at important spots.

As a representative study of the potential of RFID data, Seo (2006) proposed the process of sales (POS2) concept, which is an extension of the POS as data source for supply chain management (SCM). Seo defined the characteristics and data type of POS2 as an offline log data and derived its value on SCM practices in the apparel industry. Based on this concept, Ahn (2005) performed an experiment and proposed the real-time goods recommendation mechanism, which detects a customer's location in real-time and recommends preferred goods by means of RFID data and an intelligent agent. Somkiat, Sineenard, Akihiro, & Toshikazu (2007) embodied a shopping assistance service using personal behavior data from RFID and camera and sensor surveillance for apparel retailing.

Although related works suggest the use of RFID data to analyze customers' preferences, these works tend to focus more on generating additional sales opportunities, rather than critical decision making.

SHOPPING PROCESS IN ONLINE ENVIRONMENT

Observation and Analysis of Online Shopping Process for Shelf Management

Usually a web server registers a web log entry for every single access to the web and saves the URL requested, the IP address from which the request originated, and a time stamp (Zaiane, Xin, & Han, 1998). It is recorded as a click stream, in other words, the user's footprint on a particular website. Therefore, if customers visit and buy some product in an online retail store, the process of the purchase is stored on the retailer's server.

Since web log data includes a large amount of unnecessary information, it is important to transform web log data into an appropriate form through data preprocessing. To examine the Internet user's behavior, generally six measurements are used: hit,[1] page view,[2] visitor,[3] user,[4] duration,[5] and session.[6] As important measurement units related positively to purchase, two factors—page view and duration time—are suggested in previous research (Choi, Han, & Lee, 2004; Jun, 2002).

Online retailers have analyzed this data to extract useful information such as the user's behavior patterns. The applications of web log analysis span five areas—personalization, system improvement, site modification, business intelligence, and usage characterization (Srivastava, Cooley, Deshpande, & Tan, 2000). Within the criteria, site modification performs the change of the product catalog on the website (shelf allocation) and business intelligence covers not only assortment but also product recommendation (Ghani, 2004).

---

1. Any request from a file or a web server. A single page likely contains multiple hits as multiple image and text files are downloaded from the web server.
2. A request to load a single page of a website on the web. A page request would result from a web surfer clicking on a link on another page that points to the page in question.
3. An individual who visits a website. A visitor or unique visitor would have multiple visits.
4. The number of different people who visit a website for some time.
5. The time a visitor stays on a specific page.
6. A record of one visitor browsing a site.

Although the display capacity of online retailing is larger than offline retailing, all products cannot be placed on one page. Hence the decision of what to display on a web page is critical to online retailers. Lee, Podlaseck, Schonberg, and Hoch (2001) suggest two measurements, impression (customer views of product links) and interest (percentage of product views leading to a click), to examine the effectiveness of current product display. In the case of a product that has been exposed many times on the website but is rarely clicked, it is concluded that the product is not attractive, hence its location should be changed in the current display. Yen (2007) introduces a guideline to evaluate web page accessibility based on several structural-based accessibility models where an innovative accessibility–popularity analysis is deployed to measure and, thereby, modify a web structure. Breugelmans and Campo (2008) investigate the overall effects of in-store displays on category sales and brand market share in an online shopping context and compare the differences in effectiveness among in-store display types. As a result, first screen displays clearly have the strongest effect on market share: They benefit from their placement on the "entrance" location, central on-screen position, and direct purchase link.

Comparing Online and Offline Shopping Process

Although there are many gaps between online and offline environments, such as retail atmosphere, consumer behavior, frequently purchased items, and so on (Kim, 2007), some shopping actions are considered important in both environments. As mentioned earlier, the frequency of requests for a particular web page and its duration on the website are defined as positive purchasing factors of a product. Similarly in offline retail, the following actions are related to shelf management: browsing, trialing or fitting, and time spent (Kirkup, 2000).

To adopt the online web log analysis method in an offline environment, the shopping process for both environment cases should be reviewed. According to the previous research, online purchase steps are as follows: impression,[7] clickthrough,[8] basket placement,[9] and buying (Lee et al., 2001). In the offline case, the following five actions are generally observed: standing,[10] viewing,[11] touching,[12] carrying,[13] fitting,[14] and buying (Somkiat et al., 2007). Based on these procedures, this study matches the two environments as shown in Figure 12.1. The first step in online product impression is matched with standing and viewing as a customer looks around in front of shelves to select a product. Clickthrough is an action

7. View of the hyperlink to a web page presenting a product.
8. Clicking on the hyperlink to view the web page of the product.
9. Placement of the item in the shopping basket.
10. A consumer stops at a certain distance in front of an object.
11. Consumer looks at a certain object.
12. A consumer touches a certain object where the object does not separate from its shelf.
13. A consumer separately picks up a certain object from its shelf.
14. Consumer uses a mirror to match a certain object with his or her appearance.

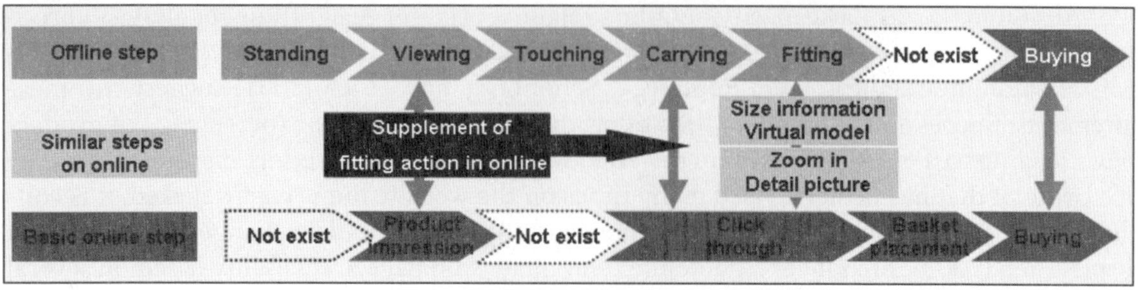

Figure 12.1. Matching of online and offline shopping steps in apparel retailing.

to choose something, hence it is like carrying an item around in offline shopping. Although the Internet environment is not the same as offline retailing, some similar functions are matched to the fitting action: zoom-in, multiple views of products on virtual model lead, picture enlargement focusing on fabric texture, and descriptions of garment performance (Kim, 2007).

RFID DATA-BASED SHELF MANAGEMENT IN APPAREL RETAILING

Design of RFID System in Apparel Retailing

To develop RFID-based shelf management, a working environment should be developed to obtain the data. Figure 12.2 shows the shopping process in apparel shopping. In this environment, the main equipment is a smart shelf, which is installed with an RFID reader since it recognizes the status of items, such as location and time on the shelf, whenever a customer picks or places an item in real time. Another important installation is the smart fitting room, which is also equipped with an RFID reader. Therefore, the smart shelf and fitting room count how many times the events are generated. Moreover, this system checks total shopping time and the duration of time from when a product leaves the shelf to when it is checked out at the counter, which is an important factor related to purchasing possibility.[15]

Based on this retail environment, the RFID system for apparel retailing is designed and displayed in Figure 12.3. The collected data from the shop floor and the back room is transmitted to an RFID middleware and stored in a local database. The headquarters gathers RFID data from all its retail stores into a data warehouse to analyze and decide each store's shelf management, and then an established shelf management plan is sent to each retail store. As this networked RFID system also links apparel purchases and the warehouse, sales

15. See http://www.envirosell.com/index.php?option=com_content&task=blogcategory&id=79&Itemid=109.

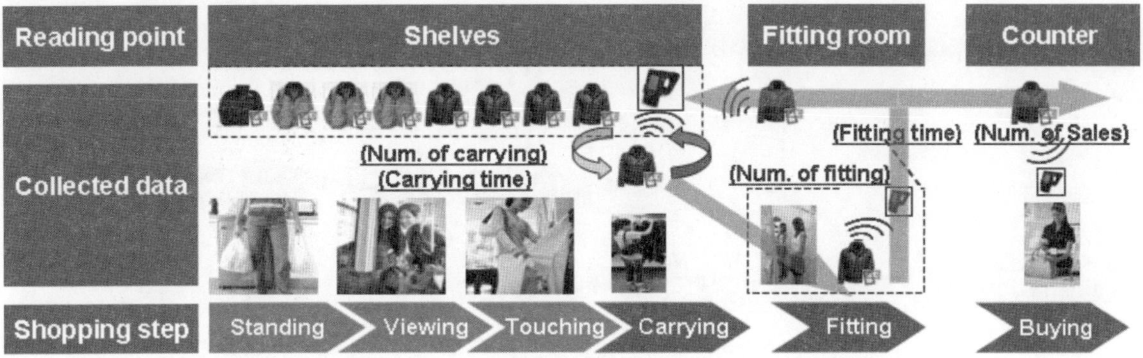

Figure 12.2. RFID data acquisition spots in apparel retailing.

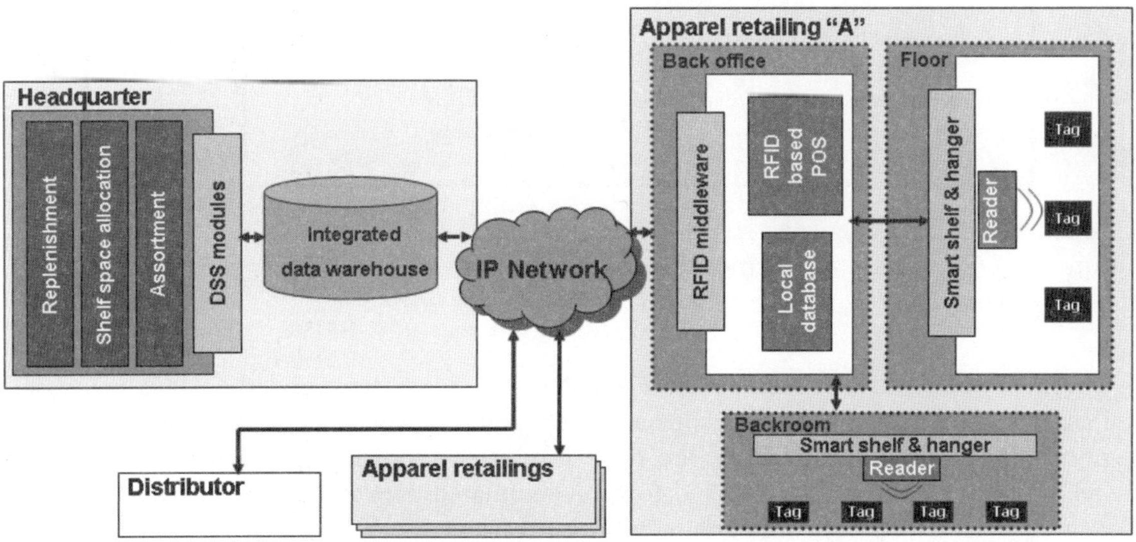

Figure 12.3. RFID system for shelf management in apparel retailing.

and inventory status is grasped in real time. Therefore, shelf management is determined more quickly and accurately.

To collect reliable data, an acquisition process shown in Figure 12.4 is suggested. In case a customer picks up an item unintentionally, it would generate meaningless data. If a carrying event is below the valid time, the carrying count data is not recorded. This process would reduce the storage overload. Fitting time begins when a customer goes into the fitting room holding an item, and it is over when the item is returned to a shelf or the POS reader transacts the payment.

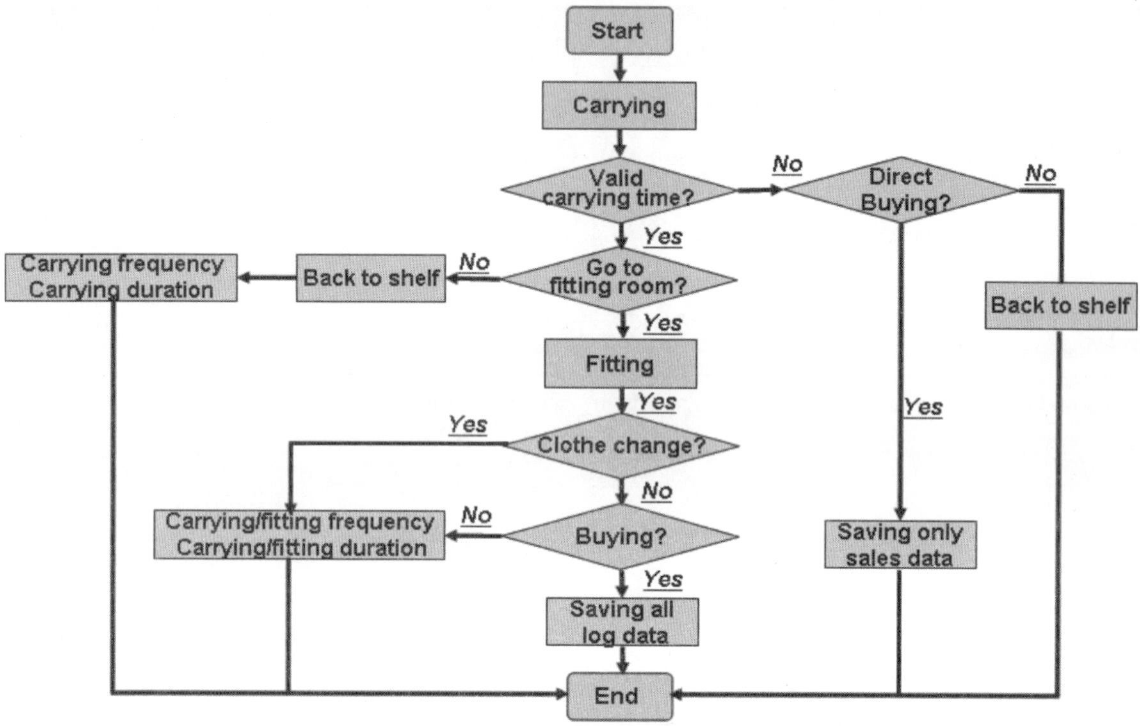

Figure 12.4. RFID data acquisition process.

The collected data is stored in the database in the system, and then it is processed for analysis. Since the RFID data format is determined on its usage, preprocessing is not required. Based on collected data, the table format is designed as shown in Figure 12.5. Instantly recorded data are converted to an accumulated data format everyday, and, in the same way, daily data are accumulated when a week is over.

Suggested Decision-Making Guidelines for Shelf Management Using RFID Data

As already mentioned, shelf management is a process that determines product arrangement on a given shelf space considering future sales. Even if the apparel headquarters establishes the plan with sophisticated data, it is difficult to meet customer demand. Generally shelf management is established based on historical sales data. Then it is implemented reflecting current sales data.

To compare the decision making based on sales data and RFID data, the scenario is as follows. Suppose a season just begins and there are two shelves, shelf A and shelf B, which display various products. Apart from differences in size and color of the products, all other

① Instantly recorded data table

Item	Carrying	Carrying time (s)	Fitting	Fitting time (s)	Sales
0001	1	5	1	53	0
0002	1	6	1	76	X
0003	1	5	0	44	0
0002	1	12	0	65	X
0004	1	7	1	51	X
0001	1	8	1	104	0

② Cumulated data table (day)

Item	Date	Carrying	Carrying time (s)	Fitting	Fitting time (s)	Sales
0001	Mon	34	231	33	697	25
0002	Mon	56	301	41	945	12
0003	Mon	5	32	1	144	1
0004	Mon	43	241	35	768	31
0005	Mon	12	69	6	320	3
0006	Mon	23	121	3	410	20

③ Cumulated data table (week)

Item	Week	Carrying	Carrying time (s)	Fitting	Fitting time (s)	Sales
0001	May 1st	234	1595	133	4896	105
0002	May 1st	356	1256	241	6841	86
0003	May 1st	435	1346	114	731	5
0004	May 1st	343	1667	335	4950	69
0005	May 1st	112	561	56	1606	15
0006	May 1st	123	820	93	2533	59

Figure 12.5. Table format of RFID data for apparel retailing.

things about the products are the same. All allocated spaces are the same for every product. For analysis, this study introduces a concept called shelf accessibility. A shelf's accessibility is determined by its distance from the entrance. For the two shelves, shelf A is closer to the entrance than shelf B. For that reason, a retailer would choose to display a promised product on shelf A rather than on shelf B, because shelf A's accessibility is higher than shelf B's accessibility.

Figure 12.6 illustrates the case of the sales data-based shelf management. It shows a scattered diagram of displayed products in terms of shelf accessibility and sales. The interpretation of the 2×2 matrix is as follows: Type 1 products are those that sell when displayed on shelves with high accessibility. Therefore, shelf space does not need to be changed. Type 3 products are satisfied beyond expectation and their location is absolutely changed to type 2 products on shelf A. They also would be moved to type 1 when their sales are higher than

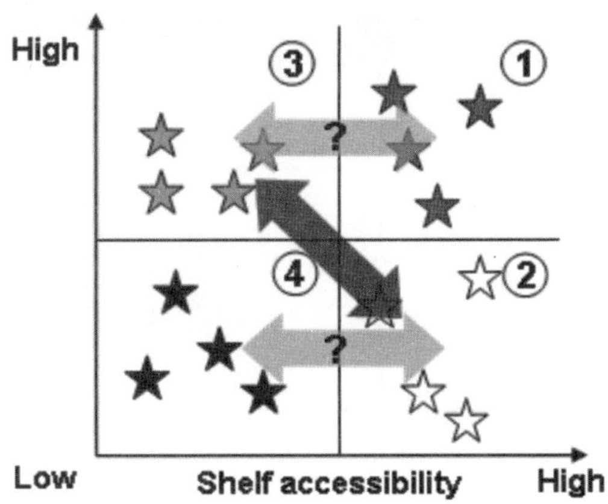

Figure 12.6. Sales data-based shelf management tool.

some type 1 products. Although type 3 products show low sales, they also change when placed on shelf A, showing higher sales than type 1. Consequently, type 2 products have to move somewhere on shelf B. However, the matrix may not guarantee higher profit since there is a possibility of misplacement. For example, some products on shelf B do not need to be moved to shelf A because enough are sold on shelf B. Or a few items on both shelves should be removed quickly since they are not likely to be sold any more. In these contexts, sales-based decision making does not provide a clear and reliable direction.

However, those problems are more specified using the RFID data as shown in Figure 12.7. To capture the preference for products on display for sale, two additional types of data from offline shopping behaviors are suggested: carrying and picking. In this analysis, three indexes are used to interpret the obtained data sequentially: carrying factor (carrying as against accessibility), fitting conversion rate (carrying to fitting), and sales conversion rate (fitting to sales).

The preconditions and analytical method are as follows. Generally, shelf accessibility and carrying count are positively correlated, hence carrying count would be increased by adjusting the product exposure. The relationships between the carrying factor (or how often an item is picked up), fitting rate, and sales rate is represented in the following equations:

When

$$Nc_i = \text{the number of times an item } (i) \text{ is picked up}$$

$$Nf_i = \text{the number of times an item } (i) \text{ is taken into the fitting room}$$

$$Ns_i = \text{the number of sales for an item } (i)$$

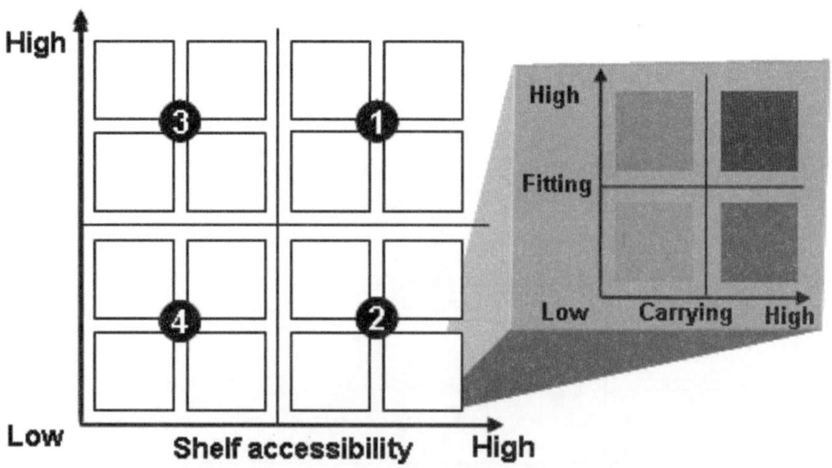

Figure 12.7. Components of RFID data-based shelf management.

then it is subject to

$$\text{N}ci \geq \text{N}fi \geq \text{N}si \; i=1, \ldots, N$$

$$\text{N}ci \geq \text{N}fi \geq \text{Nsi} \geq 0.$$

The importance of the index is in the order of carrying, fitting, and sales. Furthermore, when the degree of a conversion rate is consistent through all the shopping steps, the choice is clear, and it should be implemented first. In a case where the situation is contradictory, it is preferable to remove the negative result rather than increase the positive result. When the products are on the same shelf and the rates are the same, the decision is determined by comparing the degree of the index from carrying to sales.

By using the conversion rates, product characteristics would be defined whether they are affected by accessibility or attractiveness. Based on the casual relationships analysis among the three indexes, a product's treatments are derived.

High-Accessed Shelf Management Using RFID Data

Table 12.1. Results and Guidelines of Shelf A

Type	*Carrying factor*	*Fitting conversion*	*Sales conversion*	*Decision*	*Decision priority*
1	↑	↑	↑	Preserve	First
2	↑	↑	↓	Preserve	Second
3	↑	↓	↑	Preserve	Third
4	↑	↓	↓	Remove	Second
5	↓	↑	↑	Preserve	Fourth
6	↓	↑	↓	Preserve	Fifth
7	↓	↓	↑	Preserve	Sixth
8	↓	↓	↓	Remove	First

As a result, type 1 products are most preferable in shelf A since its high traffic is linked to high sales performance. In this situation, retailing should prepare enough inventories of those products and consider expanding its sales space. On the other hand, type 2 items sold well in spite of their high carrying factor and high fitting conversion rate. It seems that they are relatively expensive products, thus they would be preserved on the current shelf considering their markdown. Also, type 6 products are considered to have similar characteristics as type 2. Thus customers do not purchase these even though the products have a high fitting conversion rate. Without the carrying factor and fitting conversion, type

3 and 5 products show a high sales conversion rate. Therefore, they would be placed and sold on shelf A.

Type 8 products show not only a low sales conversion rate but also a low carrying factor and a low-fitting conversion rate. Although type 4 products show a high carrying factor, they failed to sell at a high rate. Those are proved to be unattractive items, and they are unlikely to be sold. In those cases, the products should be removed from the shelves completely.

Low-Accessed Shelf Management Using RFID Data

Table 12.2. Results and Guidelines of Shelf B

Type	*Carrying factor*	*Fitting conversion*	*Sales conversion*	*Decision*	*Decision priority*
1	↑	↑	↑	Move	First
2	↑	↑	↓	Preserve	First
3	↑	↓	↑	Preserve	Second
4	↑	↓	↓	Remove	Second
5	↓	↑	↑	Move	Second
6	↓	↑	↓	Preserve	Third
7	↓	↓	↑	Preserve	Fourth
8	↓	↓	↓	Remove	First

Although type 1 items are displayed on shelf B, they show not only a high carrying factor but also a high fitting conversion rate and sales conversion rate. Hence the products seem to appeal to customers, and they should be more exposed. Type 5 items on shelf B, which show high fitting and sales conversion rates without high carrying factors, also need more exposure since their sales possibility is very high once customers pick them up. Consequently, these two types of products should be moved to shelf A to increase the carrying count to achieve more sales by means of exposure. These products would be best placed in the empty space on shelf A. If some products are competing to move to shelf A, their priority is determined by actual sales quantity.

Type 3 and 7 items show high sales conversion rates in shelf B without high fitting conversion rates. Customers do not try them on frequently. Despite the relatively low customer interest, those products are selling well. Therefore, they do not need to be exposed, and it is better to keep them at their current shelf space. On the other hand, in shelf B, the sales conversion rates of type 2 and 6 products are low even though they have high fitting conversion rates. Like the earlier case, those items must be high-priced items, hence customers are hesitant to buy them. They also should remain displayed on their current shelf space for a while, and their execution priority is lower than high sales conversion rate products.

In shelf B, type 4 and 8 items show not only low sales conversion rates but also low fitting conversion rates like some products on shelf A. These also should be removed from the shelves. And new products in the store that have been waiting for release should be placed in the empty spaces on shelf B. Above all, if RFID-based shelf management is performed with a consideration to the original sales quantity, then a selected optimal case may increase the total profit.

An RFID-based shelf management guideline would be useful during the early period of a season when sales data are not enough to estimate customer demand. Even though sales data are reliable, they are not useful if the sample size is very small. In this context, RFID data would help to replace the shortage of sales data. RFID data would overcome this problem through data mining, thus ensuring its reliability. When the season starts, the weighting factor for sales is the lowest. Carrying is medium and fitting shows the highest value. However, this situation is reversed as the season progresses. When enough sales data are collected, their value becomes higher. With these data, shelf management will be more responsive and stable since sales data-based guidelines are more specified.

The guidelines of RFID-based shelf management are also used for other limited space management in apparel retailing. For example, a back room would be organized with the same assortment and same shelf space allocation ratio reflecting floor shelves' constitution using warehouse and an adjacent retail's inventory. To secure the inventory of each retail store, networked RFID systems are used to capture inventory information not only on the warehouse side but also in the adjacent retail inventory. This is done in real time to share and deliver required inventory right on time. Also, the window display, considered another shelf space, is dynamically changed with hot items.

Moreover, several important decisions are elaborated using the tool. For instance, replenishment from the warehouse to retail is fulfilled earlier for the products that show high sales conversion rates and fitting conversions.

CONCLUSION

This study proposes an RFID system for shelf management in apparel retailing to enhance sales opportunity using customer preference in a shopping process. The suggested guidelines provide insights to apparel retailers who are going to adopt RFID technology to increase the efficiency of retail operations as well as sales opportunities by specifying decision strategies compared with data-based sales guidelines.

However, some important considerations in shelf management are not covered. First, the study does not consider substitute and complementary relationships between goods, called "cross effect," that affect the decisions of assortment and shelf space allocation. Second, there are no differences in price among the products. Finally, it is not verified that these guidelines increase total profit.

Therefore, further study needs to be undertaken to provide a more detailed and integrated decision-making tool that contains cross effect and price. Moreover, an experiment based on the suggested RFID system is needed to explore the reliability of the RFID-based shelf management guidelines.

REFERENCES

Ahn, J.-M. (2005). *A study on real-time preference goods recommendation using RFID based shopping moving line analysis*. Seoul, Korea: Soongsil University.

Anderson, E. E., & Amato, H. N. (1974). A mathematical model for simultaneously determining the optimal brand-collection and display-area allocation. *Operations Research*, *22*(1), 13–21.

Borin, N., & Farris, P. (1995). A sensitivity analysis of retailer shelf management models. *Journal of Retailing*, *71*(2), 153–171.

Breugelmans, E., & Campo, K. (2008). Can in-store displays improve category sales and brand market share in online stores? A study on the overall effectiveness and differences between display types in an online FMCG context. Retrieved from http://edocs.ub.unimaas.nl/loader/file.asp?id=1344

CapturePlan. (2005). CapturePlan instead of planogram. Retrieved from http://www.corpdata.com.au/extras/pdf/whitepaper/CapturePlan%20instead%20of%20planogram.pdf

Choi, J.-H., Han, S.-M., & Lee, Y.-S. (2004). The relationship between stickiness and purchasing behavior in Internet shopping mall. *Korean Journal of Marketing*, *19*(2), 3–22.

Corstjens, M., & Doyle, P. (1983). A dynamic model for strategically allocating retail space. *Journal of the Operational Research Society*, *34*(10), 943–951.

Drèze, X., Hoch, S. J., & Purk, M. E. (1994). *Shelf management and space elasticity*. Chicago, IL: University of Chicago.

Ghani, R. (2004). Mining the web to add semantics to retail data mining. *Lecture Notes in Computer Science*, pp. 43–56.

Janiszewski, C. (1998). The influence of display characteristics on visual exploratory search behavior. *Journal of Consumer Research*, *25*(3) 290–301.

Jun, H.-C. (2002). *A study on web shopping mall user behavior using log files*. Seoul: Korea Institute of Science and Technology.

Kim, H.-J., & Kim, E.-Y. (2006). An exploratory study of RFID benefits for apparel retailing. *Journal of Korean Society of Clothing and Textiles*, *30*(12), 1697–1707.

Kim, M.-J. (2007). *Consumer perception of apparel products in Internet shopping*. Corvallis: Oregon State University.

Kirkup, M., & Carrigan, M. (2000). Video surveillance research in retailing: Ethical issues. *International Journal of Retail & Distribution Management*, *28*(11), 470–480.

Lee, J., Podlaseck, M., Schonberg, E., & Hoch, R. (2001). Visualization and analysis of clickstream data of online stores for understanding web merchandising. *Data Mining and Knowledge Discovery*, *5*(1), 59–84.

Newman, A., & Cullen, P. (2002). *Retailing: Environment & operations*. London, England: Thomson Learning Emea.

Rajaram, K. (2001). Assortment planning in fashion retailing: Methodology, application and analysis. *European Journal of Operational Research*, *129*(1), 186–208.

Rao, B. P. (2000). Improving retail effectiveness through technology: A survey of analytical tools for physical and on-line retailers. *Technology in Society*, *22*(1), 111–122.

Seo, J.-H. (2006). *A study on the value of RFID based process of sales (Pos2) system in SCM: Insight from case study of apparel industry*. Daejeon, Korea: Information and Communications University.

Somkiat, S.-U., Sineenard, P., Akihiro, O., & Toshikazu, K. (2007). Consumer-friendly shopping assistance by personal behavior log analysis on ubiquitous shop space. In *Proceedings of the 2nd IEEE Asia-Pacific Service Computing Conference* (pp. 496–503). Washington, DC: IEEE Computer Society.

Srivastava, J., Cooley, R., Deshpande, M., & Tan, P. N. (2000). web usage mining: Discovery and applications of usage patterns from web data. *SIGKDD Explorations*, *1*(2), 12–23.

Urban, T. L. (1998). An inventory-theoretic approach to product assortment and shelf-space allocation. *Journal of Retailing*, *74*(1), 15–35.

Wessel, Rhea. (2007, September 20). Metro group's Galeria Kauhof launches UHF item-level pilot. *RFID Journal*. Retrieved from http://www.rfidjournal.com/article/articleview/3624/1/1/

Yen, B. P. C. (2007). The design and evaluation of accessibility on web navigation. *Decision Support Systems*, *42*(4), 2219–2235.

Zaiane, O. R., Xin, M., & Han, J. (1998). Discovering web access patterns and trends by applying OLAP and data mining technology on web logs. In *Proceedings of the Advances in Digital Libraries Conference* (pp. 19–29). Washington, DC: IEEE Computer Society.

CHAPTER 13

PRODUCTIVITY IMPROVEMENTS IN RESOURCE-CONSTRAINED ENVIRONMENTS: A SYSTEMATIC FRAMEWORK

Anil Mital

Department of Mechanical Engineering
University of Cincinnati

Arunkumar Pennathur

University of Texas at El Paso

Rajeev Senapati

University of Texas

INTRODUCTION

Manufacturing plays a vital role in the economic health of any nation. In recent years, manufacturing has consistently generated over about one fifth of the U.S. gross national product. According to the *Statistical Abstract of the United States* (1997), manufacturing companies in the United States employ almost 19 million workers and add $1.7 trillion in value to the U.S. economy. As a result of globalization of markets, most companies are faced with an urgent need to manufacture high-quality products at a lower cost and to make them available faster than ever before. Constantly changing market situations also require quick and flexible reactions on the part of manufacturers and a reduction in the lead time for product development and manufacture. The international marketplace requires excellence in the manufacturing, placing new emphasis on *quality*, *service*, and *lower costs*.

In the early 1980s, manufacturing researchers and practitioners advocated completely automated manufacturing factories as the shape of the future factory (e.g., Bullinger & Warnecke, 1985). The workerless factories were expected to provide a way to deal with the productivity woes and prompted U.S. manufacturers to invest billions of dollars in high-technology equipment automation. These investments, however, have been without matching payoffs in the form of higher productivity. Recent data show that U.S. productivity has slowed down significantly since 1973 (Sparks & Greiner, 1997). It is now being acknowledged, albeit reluctantly, that completely automated factories are not a reality, at least in the foreseeable future, due to a variety of technical, economical, and cybernetic reasons (Mital, 1997) and cannot lead to higher productivity. It is also being realized that humans will continue to play an essential and integral role in modern manufacturing. However, since humans are error-prone, they can be a major cause of lower productivity (the cause of error can inherently be the operator or the situation). Compared to 10 other industrialized countries, as well as Canada, Japan, and eight other newly industrialized nations, U.S. labor productivity faired poorer from the years 1979 to 1995. U.S. manufacturing labor productivity (output per hour), for instance, grew only 2% in 2005, down from a 6.1% increase in 2004 (Figure 13.1). This marginal increase in output placed the United States only fourth among 17 competing economies. The largest output increases were about 9% in South Korea and 7% in Sweden (Figure 13.2).

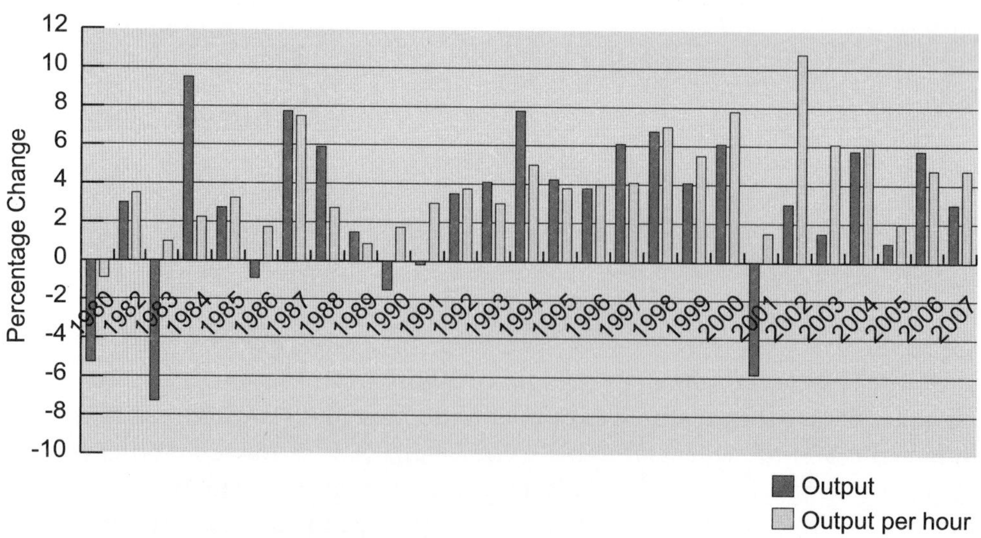

Figure 13.1. Annual percentage change in manufacturing productivity in the United States, 1980–2007.

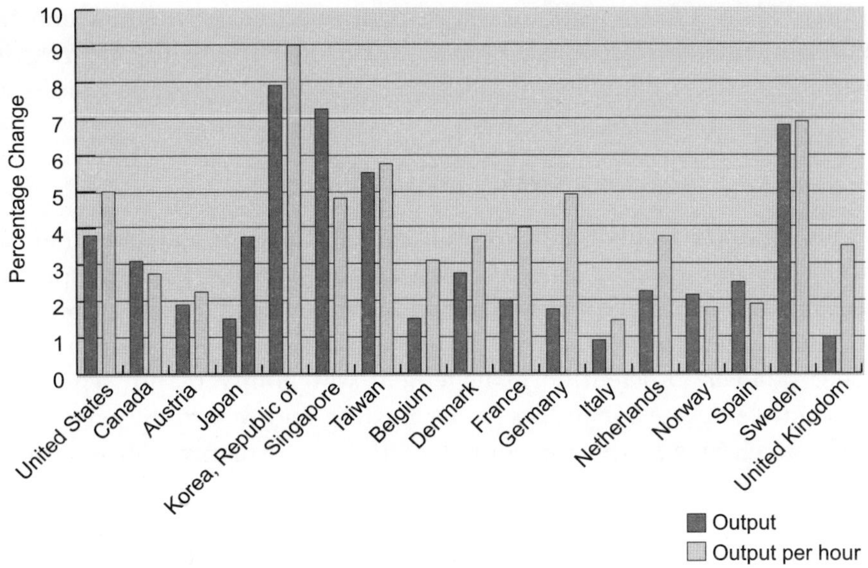

Figure 13.2. Annual percentage change in manufacturing productivity in 17 competing countries, 1993–2007.

It is imperative that in the face of international competition, increases in U.S. productivity continue if the nation is to maintain a high standard of living and generate economic growth. This means that manufacturers must constantly look for ways and means to enhance their productivity. One way to achieve higher manufacturing productivity is to identify and measure the specific causes of productivity loss and mitigate them. Problems such as worker injuries, frequent machine breakdowns, large amount of scrap generated, and so on lower productivity. However, these problems are all too apparent in a plant, and corrective action can easily be taken to correct, eliminate, or reduce the extent of the problem. The situation becomes more challenging once the obvious and very visible sources of productivity impediments are mitigated. In the absence of any visible signs of productivity problems or opportunities for productivity improvement, the task of enhancing productivity becomes critical. The situation is further complicated as very few manufacturing firms have effective productivity measurement systems that can pinpoint the specific reasons for changes in productivity relationships (Gold, 1982). If such opportunities can be identified and evaluated, it may be possible to develop and use assistive devices (broadly defined as any device that assists the operator in enhancing his or her job performance) to enhance productivity.

The objective of this chapter is twofold: (a) to develop a conceptual framework for identifying productivity enhancement opportunities in manufacturing industry when such opportunities are not readily apparent and (b) to present a case to illustrate the implementation of the proposed framework. This chapter is organized as follows: the second section briefly reviews productivity-related literature with an emphasis on the productivity measurement techniques and models. A framework for a systematic identification of productivity improvement potential in manufacturing is presented in the third section. The fourth section presents a case of a parking systems manufacturer, illustrating the implementation of the framework.

BACKGROUND

Since the beginning of the Industrial Revolution, practitioners, researchers, and policymakers have tried to define and evaluate particular workplace practices that can spur productivity growth and competitiveness. The organized approach to productivity improvement in the American industry began with Frederick Taylor's scientific management principle (Taylor, 1915). Another pioneer in scientific management was Henry Gantt, who developed the Gantt chart, a simple device for precisely anticipating and recording performance (Gantt, 1974). The Gantt chart was the forerunner of the Performance Evaluation and Review Technique (PERT) charts. Lillian and Frank Gilbreth were two other industrial engineering pioneers who carefully studied peoples' movements as they worked. They broke jobs down into very small elements so that each movement could be analyzed and perhaps simplified or eliminated (Gilbreth & Gilbreth, 1973).

Productivity is now, in general, regarded as the relationship between the outputs generated from a system and the inputs provided to create those outputs (Sink, 1991). Specifically, productivity is defined as the relationship of the amount produced by a given system during a given period and the quantity of resources consumed to create or produce these outputs over the same period. Typical inputs are in the form of labor (human resources), capital (physical and financial capital assets), energy, materials, and data. These resources are then transformed into outputs (goods and services). This relationship is illustrated in Figure 13.3. Different techniques exist to measure the costs associated with these outputs and inputs (Stewart, 1983).

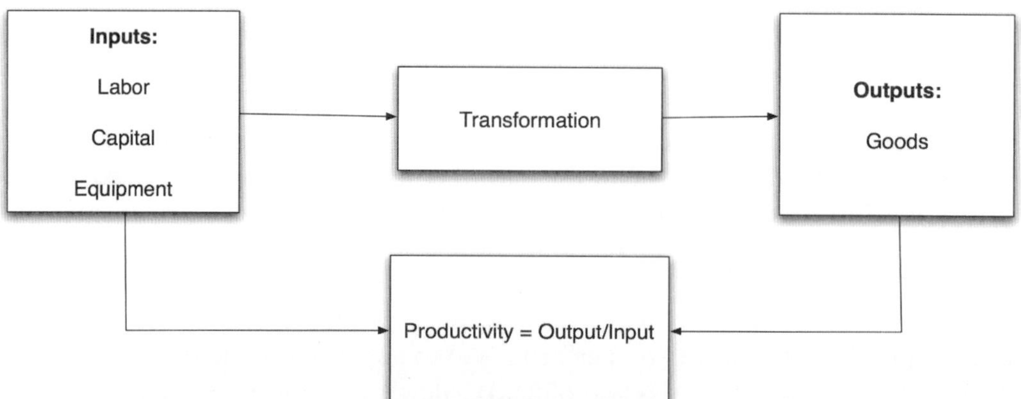

Figure 13.3. General productivity concept.
Source: Modified from Sink (1985).

In general, two forms of productivity are reported in practice: partial productivity and multifactor productivity. Partial productivity is the ratio of gross or net output to just one type of input from among different inputs, as seen in Figure 13.3. Examples of partial productivity are labor productivity and capital productivity. Total or multifactor productivity measures, on the other hand, compare all outputs with all inputs. Among the important measures for partial and multifactor productivity, the Tornqvist index is widely used in the U.S. Bureau of Labor Statistics (BLS) for reporting partial and multifactor productivities. For example, the labor productivity (partial), according to the Tornqvist index (Caves, Christessen, & Diewert, 1982) for industries that produce a number of different products or provide a number of different services, is given by

$$\frac{Q_t}{Q_{t-1}} = \exp\left[\sum_{i=1}^{n} w_{i,t}\left(\ln\frac{q_{i,t}}{q_{i,t-1}}\right)\right] \qquad (1)$$

where

$\frac{Q_t}{Q_{t-1}}$ = the ratio of output in the current year (t) to the previous year (t-1)

n = number of products

$\ln\frac{q_{i,t}}{q_{i,t-1}}$ = the natural log of the ratio of the quantity of product i in the current year to the quantity in the previous year

$w_{i,t}$ = the average value share weight for product i

This measure of labor productivity relates only to one input, namely, labor time. The measure is also a reflection of a number of interrelated influences on product manufacturing, such as technology change, capital investment per worker, capacity utilization, layout and flow of material, skills and effort of the workforce, managerial skill, and labor management relations, among others (BLS, 1997).

BLS uses a number of sources of information on the output and input used in the calculation of partial and multifactor productivities. Outputs are developed from the Bureau of Census's information, as a deflated value of production or physical quantity of production in a certain industry. Inputs are specific to the type of partial productivity: Labor input is obtained by dividing the aggregate employee hours for each year by the base period aggregate. Capital input is based on the flow of services derived from existing stock of physical capital, including equipment, structures, land, and inventories. Intermediate purchases include real materials, services, fuels, and energy and electricity consumed by the industry (BLS, 1997). The Tornqvist index for multifactor productivity is given by

$$\ln\left(\frac{A_t}{A_{t-1}}\right) = \ln\left(\frac{Q_t}{Q_{t-1}}\right) - \left[w_k\left(\ln\frac{K_t}{K_{t-1}}\right) + w_l\left(\ln\frac{L_t}{L_{t-1}}\right) + w_{ip}\left(\ln\frac{IP_t}{IP_{t-1}}\right)\right] \quad (2)$$

where

ln = the natural logarithm of the variable
A = multifactor productivity
Q = output
K = capital input
L = labor input
IP = intermediate purchase input
w_k, w_l, w_{ip} = cost share weights

The weights are the means of the cost shares in two adjoining time periods.

Matta (1989) prescribed a weighted multifactor productivity model, where the total productivity for a given firm is a function of various partial productivities. The model is based on the following decision-making criteria: (a) potential for improvement by the plant manager and (b) the relative importance of organizational and management goals.

Performance measures indicate how well an organization utilizes its labor, capital, and raw material resources among other inputs. The ability to compete effectively in both domestic and international markets has become the most important business issue of the 1990s (Hamel & Prahalad, 1994). A major study conducted by Queens University in Canada focused on identifying manufacturing practices that have a significant influence on company success (Sohal, Gordon, Fuller, & Simon, 1999). Data were collected by means of a postal questionnaire that was mailed to senior staff members in 5,000 manufacturing industries. Improved quality and time reduction were the major factors identified for productivity improvement by this survey.

These findings are in agreement with what other experts consider measures of productivity and performance: quality, cycle time, and customer satisfaction (American Productivity Center, 1981; Craig & Harris, 1973; Miller, 1992; Mohanty & Rajput, 1988; Sandman & Hayes, 1980; Sumanth, 1981). Each of these elements of performance has limited value when viewed independently. In isolation none of them can fully measure performance or fully describe how well the organization is doing. For example, high levels of productivity would not be meaningful if cycle times were increasing or customer service level was dropping. Each of the primary performance measurements must be considered in tandem when judging the total activity performance. A manufacturing firm's competitive advantage in the market is a function of the plant's ability to (a) produce high-quality products, (b) provide superior delivery performance, and (c) be a low-cost producer (Umble, 1988). The degree to which a manufacturing firm can achieve these three objectives largely determines its long-term financial success.

As this brief review of productivity literature shows, existing productivity and accounting measures need improvement so that manufacturing performance in a setting where human input in work is essential can be measured more accurately. It is also clear that the key mix of measurements of productivity must be derived from quality, cost, cycle time, and customer satisfaction. A family of measures such as this, rather than one isolated measure, will enable the analyst to view the organization as a whole, single, complex, and dynamic system. Also most of the productivity measurement and improvement efforts in the past have been more reactive in nature when the productivity problems are readily apparent and obvious to the analyst. A proactive approach should be adopted to identify opportunities for improvement even though there are no visible signs of productivity problems. There is a need to develop a systematic way to identify these opportunities, prioritize them according to their importance, and find ways to enhance the effectiveness of the human element contributing to the potential for improvement. The next section presents a model to systematically identify areas for productivity improvement.

THE MODEL

Figure 13.4 shows the proposed productivity improvement model. The model is driven by the company's productivity goals. If the goals are not met, there is a need to improve the productivity of the system.

Step 1: Identifying Critical Activities

The first step in the model is to identify areas that have maximum potential for improvement. Key factors used for this identification of critical activities are time, cost, and quality.

Determination of Time-Critical Activities

It is possible to reduce the time of any activity by deploying better techniques and equipments, but the real benefit will occur only if the processing time on bottleneck activities is reduced, which in turn reduces the cycle time. Critical path analysis (CPM) is a very effective technique used for identification of bottleneck activities. For more information on the CPM technique, refer to Radcliffe, Kawal, and Stephenson (1967) and Shaffer, Ritter, and Meyer (1965). Figure 13.5 shows the CPM process.

Identification of Cost-Centered Activities

Attention must be paid to activities where the maximum cost or value of the product is centered. Activity-based costing (ABC) is a technique that identifies the relationship between an activity and the cost associated with it. Placing costs on activities and their outputs provides a clear metric for determining improvement priorities. It is through ABC that an organization can begin to see actual dollar cost against individual activities and find opportunities to streamline or reduce the costs (Figure 13.6). Readers should refer to O'Guin (1991) for more information on ABC.

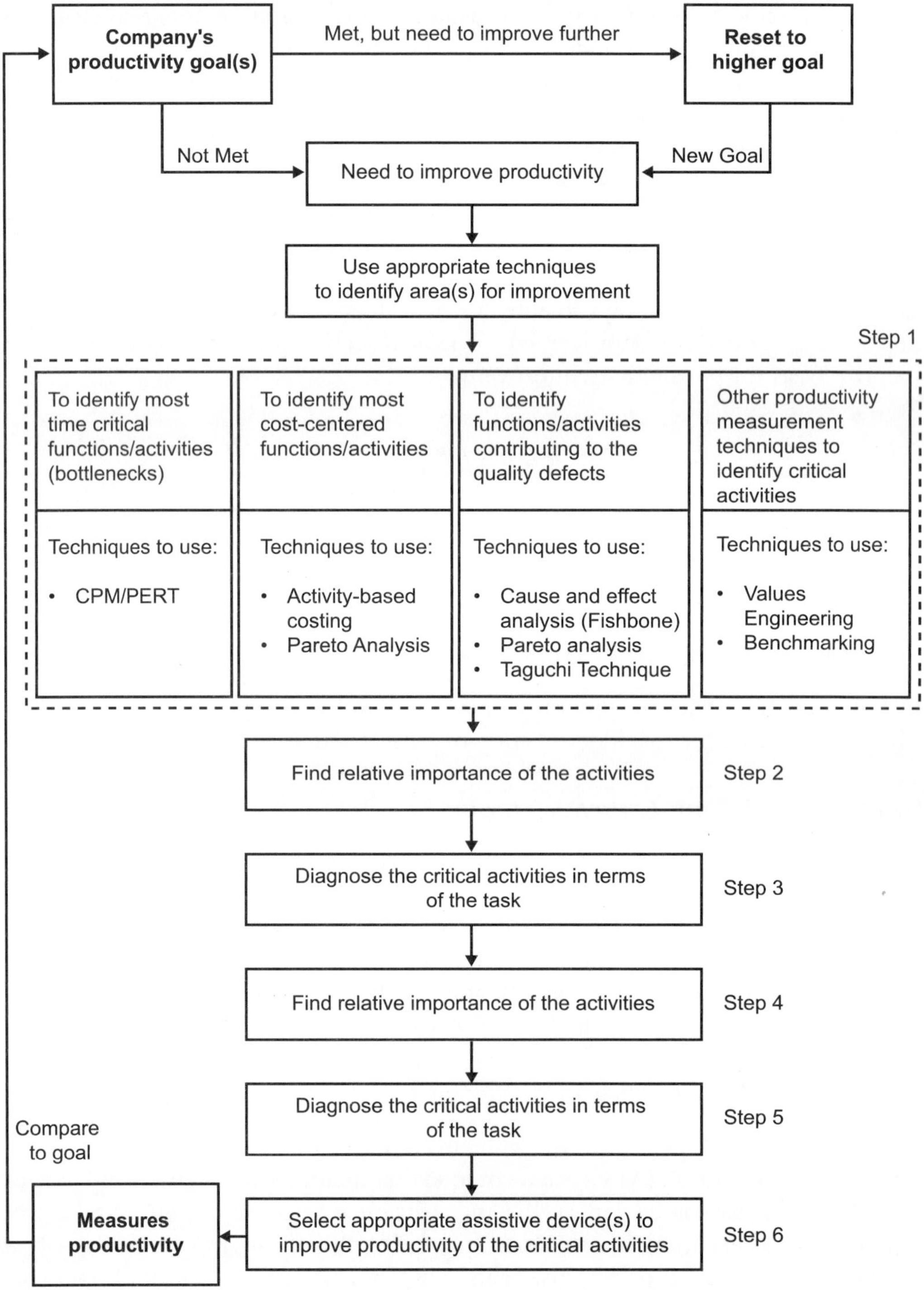

Figure 13.4. Conceptual framework for productivity improvement process.

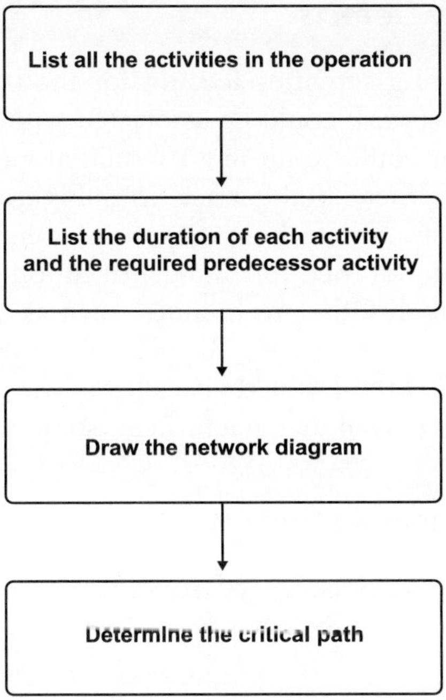

Figure 13.5. CPM process.

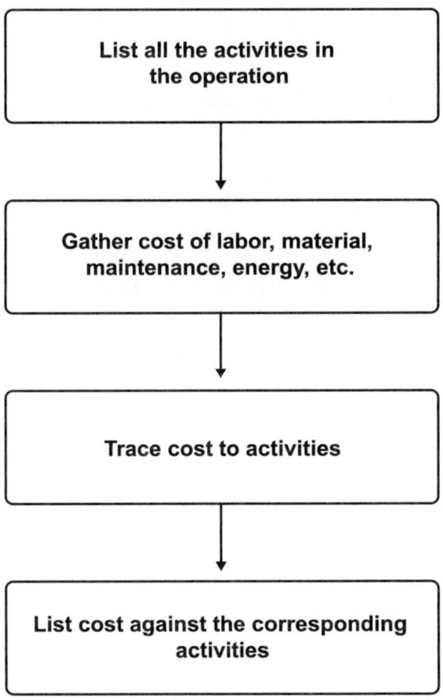

Figure 13.6. Activity-based costing process.

Determination of Quality Defects

Often, few of the manufacturing activities account for the bulk of quality problems. For example, 80% of the cost of rework could be associated with 20% of the activities. Thus, improving just these few areas could result in a substantial increase in productivity. Pareto analysis can be used to identify these vital few activities. Figure 13.7 shows tasks involved in carrying out the Pareto analysis. The techniques presented in this section do not represent a comprehensive list of techniques used for the identification of areas with maximum productivity improvement potential. Other techniques, such as quality function deployment, value engineering, cause and effect analysis, Taguchi methods, benchmarking, and so on, are discussed by Kansal (1999). Step 1 would give a list of those activities that are critical to improving the productivity of a given manufacturing system.

Step 2: Prioritizing Critical Activities

The next step in our model is to prioritize activities identified in step 1 based on their relative importance in order to select those activities that can maximize productivity and help in

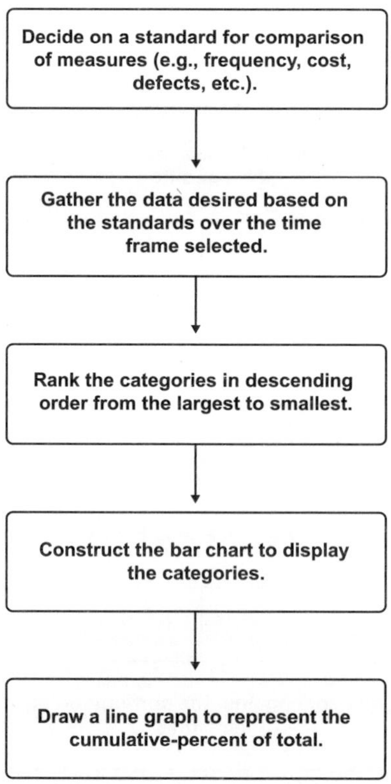

Figure 13.7. Pareto analysis to determine vital few activities.

achieving the company's goal. A weighted-ranking technique is used to rank those activities. The weight of each criteria, time, cost, and quality, is calculated as the proportion of the value that the criteria adds to the activity.

The time weight for an activity is the ratio of time spent on the activity to the total time spent on the product.

$$w_{tA} = \frac{T_A}{T_{Total}} \quad (3)$$

where

w_{tA} = time weight for activity A
T_A = time spent on activity A
T_{Total} = total time spent on the product

The cost weight for an activity is the ratio of the cost associated with activity A to the total cost of the product.

$$w_{cA} = \frac{C_A}{C_{Total}} \quad (4)$$

where

w_{cA} = cost weight for activity A
C_A = cost (labor + material) for activity A
C_{Total} = total cost (labor +material) for the product

Quality weight is the ratio of the cost of rework, warranty, scrap, and so on due to a defect on the activity to the total cost of rework, warranty claims, scrap, and so on for the product.

$$w_{qA} = \frac{Q_A}{Q_{Total}} \quad (5)$$

where

w_{qA} = quality weight for activity A
Q_A = cost of rework, warranties, and so on to activity A
Q_{Total} = total cost of rework, warranties, and so on for the product

Add the three weights for each activity identified from step 1. Activities with the highest weight are the ones consuming most of the resources and hence have the potential for maximum savings. These activities are selected for further analysis to pinpoint the opportunities for improvement.

Step 3: Diagnosing Critical Activities

Performance of a manufacturing activity depends on four variables: (a) the task performed, (b) the operator(s) performing the task, (c) the materials or tools used for the task, and (d) the environment in which the task is carried out. Each of these variables affects the performance of an activity in a different way. Often, poor system performance is equated with poor human performance without taking other variables into account. Human performance may be degraded because of poor design decisions, including the tools being used for the task. The environment in which people work also influences them and, at times, may enhance or degrade their performance. Poorly designed systems do not elicit optimum performance from workers. It is possible to enhance performance, efficiency, and productivity of the workers given the right kind of assistive aids and the right environment.

Supposing, in a manufacturing operation, inspection activity is found to have the highest potential for productivity improvement (from steps 1 and 2 of the analysis). The activity is broken down into task, operator, machine or tool, and environmental variables. Sample hierarchical analysis of the inspection function in manufacturing is shown in Figure 13.8. It is found that the inspection activity consists of a task of measuring the dimensions and hardness of the objects and of comparing the test surface to a standard surface finish. The comparison of the surface finish is found to be the most difficult and time-consuming activity. Judgment by visual inspection alone is also highly unreliable. The operator has to perform the activity in the standing position. Objects to be inspected are very small, and the surface of the object contrasts sharply with the background. In real life, an activity may have several variables, in which case they must all be listed systematically under each category.

Step 4: Selecting Appropriate Assistive Devices

The final step is to identify an appropriate assistive device or devices to enhance productivity of the activity variables: task, operator, machine, and environment. A surface comparator can

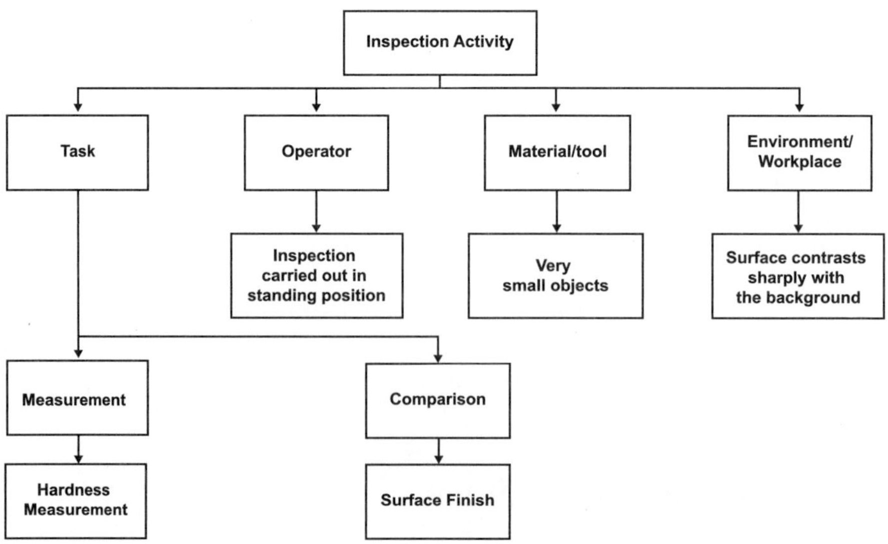

Figure 13.8. Sample hierarchical analysis of the inspection activity in manufacturing.

be an effective assistive device to aid in the comparison of the two surfaces. Operators can be provided with a sit/stand stool to relieve him of pain. If it is not possible, provide a composite floor of recycled tires to relieve strain and stress on the back. If the objects to be inspected are very small, proper magnification can be provided to increase the accuracy and efficiency of the inspection operation. Efficient illumination is very essential for the inspection task. For a surface that contrasts sharply with the background, front lighting should be used.

Thus, the task, operator, environment, and material variables should be systematically analyzed to improve the performance of the activity. One can compile a list of other assistive devices in assembly and inspection areas for this purpose (Kansal, 1999).

CASE STUDY

Introduction to the Case Study

A case study was performed at Amano Inc. to show the implementation of the proposed productivity model. Amano is engaged in the development, manufacturing, and distribution of time recorders, time and attendance systems, and parking control systems. The product chosen was the Fee Indicator, which is used to display parking fees and change due along with the current time as communicated by the fee computer. Figure 13.9 shows the Fee Indicator.

Data Collection

Data on manufacturing process, activity times, equipment or tools used, and so on was obtained from company records and discussion with the engineer and the operator of the device. Most of the components are purchased from the vendors, and thus the major activity of the company is the assembly of the components. The company manufactures 8 of these indicators in a day but would like to increase the capacity to 10 a day. Only one operator is assigned to this product. For accounting purposes, the operating costs are taken as $60.00 per hour. This includes the labor cost of $12.70 per hour and $47.30 per hour for plant maintenance. Table 13.1 lists all the major activities for the Fee Indicator and their associated time durations and costs.

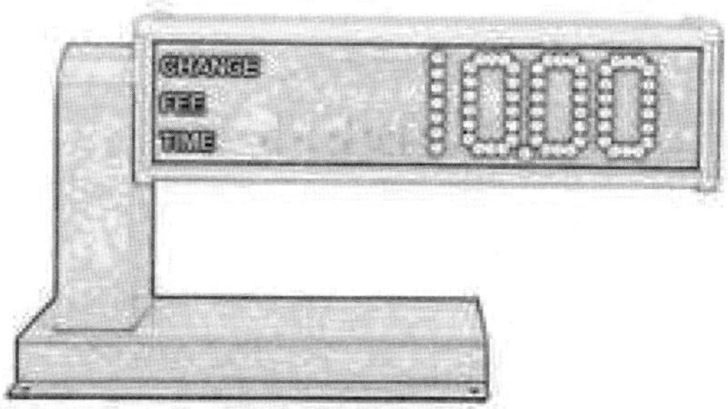

Figure 13.9. The Fee Indicator.

Table 13.1. Major Activities for the Fee Indicator and Their Associated Time Duration and Costs

Activity	Description	Duration (min)	Operating cost @ $60/hr	Material cost ($)	Total cost ($)
A	Terminal box cover assembly	1.0	1.0	9.80	10.80
B	End cap assembly	1.28	1.28	15.75	17.03
C	Display panel assembly	8.66	8.66	120.88	129.54
D	Lens/extrusion/display assembly	4.07	4.07	22.58	26.65
E	Terminal box to base assembly	5.57	5.57	15.47	21.04
F	Display/base assembly	17.42	17.42	38.00	55.42
G	Final inspection/package	16.45	16.45	0	16.45
	Total	54.45	54.45	222.48	276.93

Model Implementation

Step 1: Identification of Critical Activities

Activities consuming maximum time. The activities are performed sequentially from A to G by one operator. In effect, every activity is a bottleneck activity. However, we can consider the two activities, which consume almost 65% of the cycle time, as time-critical. These activities are (a) display and base assembly and (b) final inspection and packaging.

Activity consuming maximum costs. Activity-based costing, shown in Table 13.2, traces all the costs associated with each activity. The Pareto analysis was performed on activity costs to find the cost-centered activities. Of the seven activities, over 65% of the cost is due to two activities: (a) display panel assembly and (b) display and base assembly. Hence, these two activities are chosen as cost-centered activities.

Step 2: Prioritizing Critical Activities

Important activities identified in step 1 are the following:

Table 13.2. Weights for Critical Activities

Activity category	Activity	Time weight	Cost weight	Total
A	Display panel assembly	0.16	0.47	0.63
B	Display/base assembly	0.32	0.20	0.52
C	Final inspection/package	0.30	0.06	0.36

- Display and base assembly
- Final inspection and packaging
- Display panel assembly

For each of these activities, the time weights and the cost weights are calculated. Activity A, in other words, display panel assembly, for example, takes 8.66 minutes out of total 54.45 minutes required for the complete product. Hence, the time weight for activity A was calculated as the following:

$$w_{tA} = \frac{T_A}{T_{Total}} = \frac{8.66}{54.45} = 0.16 \qquad (5)$$

Similarly, the cost associated with activity A is \$129.58 out of a total cost of \$276.93. Hence, the cost weight for activity A was calculated as the following:

$$w_{cA} = \frac{C_A}{C_{Total}} = \frac{129.58}{276.93} = 0.47 \qquad (6)$$

In a similar manner, weights for the other two activities were calculated, and the results are presented in Table 13.2. The company kept no records of the defects, warranties, claims, and so on. Hence, quality-related weights are not taken into consideration.

Display panel assembly, which has the highest total weight, represents the maximum potential for improvement. Depending on the productivity goal of the organization and the resources available, one or more of the critical activities may be chosen for improvement. In the present case, display panel assembly was selected for further analysis and improvement.

Step 3: Diagnosing Critical Activity

The activity selected from step 2 was analyzed for the four activity performance variables: task, operator, tools, and environment or workplace. Table 13.3 displays the important tasks involved in the activity and the tools used. Task, tools or materials, and operator variables are listed in Table 13.4.

Step 4: Identification of Assistive Devices

Once the activity was broken down into task, operator, machine, and environment or workplace variables, it was possible to suggest assistive device(s) that can aid in enhancing the performance of some or all of the variables that, in turn, would increase the productivity of the activity. Suggested assistive devices for each activity variable are listed in Table 13.5.

If the productivity goal of the organization is not met after implementing the assistive devices for display panel assembly, then the second-most critical activity, display and base

Table 13.3. Tasks Involved in Display Panel Assembly and the Tools Used

Task	*Tool*
Pick circuit board from bin	
Cut mask	Scissors
Pick chip from bench	
Assemble chip to PCB	
Pick screws from bench	
Pick washers from bench	
Assemble circuit board to panel	Screwdriver
Assemble power supply to panel	Screwdriver
Assemble cables to circuit board	Screwdriver
Aside assembly to cart	

Table 13.4. List of Variables

Task variables	Components like chip and screws are picked up from the bench or bins
	Completed assembly is carried to the testing area by cart
Operator variables	Operator has to stand for the assembly task
	Operator neglects to install a chip or inserts a wrong chip
	Operator takes longer breaks
Tools/materials variables	Manual screwdriver is used
	Parts at the station are not of correct type or in right quantity

assembly, is taken for improvement, and steps 3 and 4 are repeated. This process is repeated until the company's goal is achieved.

CONCLUSION

This chapter describes a systematic procedure to (a) identify improvement areas in a manufacturing setup, (b) rank and prioritize those activities that have maximum potential for improvement, and (c) identify assistive devices to enhance productivity. Techniques such as CPM or PERT, activity-based costing, Pareto analysis, and quality function deployment can be used very effectively to identify critical activities in manufacturing operations. This multicriteria model is based on three key criteria to measure productivity: time, cost, and quality. A real-life case study performed on a typical manufacturing company confirms the validity of the model.

Table 13.5. List of Suggested Assistive Devices for the Display Panel Assembly

No.	Assistive devices
1	For the ease of discharging by hand, use grab tongues and tipping devices. These devices discharge one component at a time at the touch of a finger. Alternatively, magazines, which have separate sections for storing different parts, can ease the task of sorting and picking the correct part.
2	Currently, completed assembly is carried to the testing area by cart. If there is any defect, it is again carried back to the assembly area for rework. This causes wasted movement and time. Instead, the testing area can be located next to the assembly area. This would reduce the time for travel.
3	The seat provided is not adequate for the sit/stand type of operation. Sit/stand stools can be provided for this purpose, which can reduce the strain from standing.
4	Operator can be provided with a step-by-step procedure to perform the assembly operation. This will reduce the possibility of errors and will also reduce the mental workload of the operator, since he no longer has to memorize the sequence of operation or the part number.
5	Close monitoring must be performed on the amount of time that the operator spends during each of the three breaks permitted during the day. Checks must also be made during the start and end of the shift. Even a 5-minute delay on these activities can accumulate to a large amount of unproductive time over a day.
6	The manual screwdriver used for the display panel assembly takes about 3 minutes out of the total of 8.66 minutes required for the activity, or about 35% of the time. Using powered screwdrivers can significantly reduce time.
7	A data management system can be developed such that the inventory person is informed of an order as soon as it is placed at the assembly. It can simply be a requisition request card, which is sent to the inventory. The inventory person can stock the workstation with sufficient pats and also monitor the status of the inventory and take corrective action.

REFERENCES

American Productivity Center. (1981). *How to measure productivity at the firm level*. Houston, TX: American Productivity Center.

Bullinger, H. J., & Warnecke, H. J. (1985). *Towards the factory of the future*. Heidelberg, Germany: Springer-Verlag.

Bureau of Labor Statistics. (1997). Industry productivity measures. In *BLS handbook of methods*. Washington, DC: U.S. Government Printing Office.

Burnham, D. C. (1992). Productivity: An overview. In G. Salvendy (Ed.), *Handbook of industrial engineering* (2nd ed.). New York, NY: Wiley.

Caves, D. W., Christessen, L., & Diewert, W. (1982). The economic theory of index numbers and the measurement of input, output, and productivity. *Ecomometrica*, *50*(6), 1393–1414.

Craig, C. E., & Harris, R. C. (1973). Total productivity measurement at the film level. *Sloan Management Review*, *14*(3), 13–28.

Gantt, H. L. (1974). *Organizing for work*. Easton, PA: Hive Publications.

Gilbreth, F., & Gilbreth, L. (1973). *Applied motion study: A collection of papers on the efficient method to industrial preparedness*. Easton, PA: Hive Publications.

Gold, B. (1982). Practical productivity analysis for management: Part 1. Analytic framework. *IIE Transactions*, *14*(4), 227–242.

Hamel, G., & Prahalad, C. K. (1994). Competing for the future. *Harvard Business Review*, 72(4), 122–132.

Kansal, A. (1999). *A systematic approach to identifying and improving productivity of critical areas in manufacturing system*. MS thesis, Department of Industrial Engineering, University of Cincinnati.

Matta, K. (1989). A goal-oriented productivity index for manufacturing systems. *International Journal of Operations and Production Management*, *9*(4), 66–76.

Miller, J. (1992). The new activity performance measures. *CMA Magazine*, *34*.

Miterek, S., Dooley, K., & Anderson, J. (1992). Productivity as a performance measure. *International Journal of Operations and Production Management*, *12*(1), 29–45.

Mital, A. (1997). What role for humans in computer integrated manufacturing? *International Journal of Computer Integrated Manufacturing*, *10*(1), 190–198.

Mohanty, R. P., & Rajput, I. (1988). Productivity measurement in a manufacturing company. *International Journal of Operations and Production Management*, *8*(4), 65–78.

O'Guin, M. (1991). *Complete guide to activity-based costing*. Englewood Cliffs, NJ: Prentice Hall.

Radcliffe, B., Kawal, D., & Stephenson, R. (1967). *Critical path method*. Chicago, IL: Cahners.

Sandman, W. E., & Hayes, J. P. (1980). *How to win productivity in manufacturing*. New York, NY: AMACOM.

Sink, D. S. (1991). The role of measurement in achieving world class quality and productivity management. *Industrial Engineering*, *23*(6), 23–28.

Shaffer, L. R., Ritter, J., & Meyer, W. (1965). *The critical path method*. New York, NY: McGraw-Hill.

Sohal, A., Gordon, J., Fuller, G., & Simon, A. (1999). Manufacturing practices and comparative capability: An Australian study. *Technovation*, *19*(5), 295–304.

Sparks, C., Gordon, J., Fuller, G., & Simon, A. (1999). *U.S. and foreign productivity and unit labor costs*. Monthly Labor Review, Bureau of Labor Statistics, Washington, DC, pp. 26–35.

Stewart, W. (1983). Multiple input productivity measurement of production systems. *International Journal of Production Research*, *21*(5), 745–753.

Sumanth, D. (1981). Productivity indicators used by major U.S. manufacturing companies: The result of a survey. *Industrial Engineering*, *5*, 70–73.

Taylor, F. (1915). *The principles of scientific management*. New York, NY: Harper & Brothers.

Umble, M. (1988). How U.S. manufacturing can regain the competitive edge. *Baylor Business Review*, Fall 1988, 2–8.

U.S. Department of Commerce. (1997). *Statistical Abstract of the United States*. Washington, DC: Bureau of the Census.

Yamashita, T. (1987). The interaction between man and robot in high-technology industries. In K. Noro (Ed.), *Occupation health and safety in automation and robotics* (pp. 139–142). London, England: Taylor & Francis.

CHAPTER 14

IMPLEMENTATION OF LEAN SIGMA METHODOLOGY FOR MINIMIZING WASTE OF NATURAL RESOURCES

Adan Valles-Chavez

Jaime Sanchez

The Technological Institute of Ciudad Juarez
Chihuahua, Mexico

INTRODUCTION

In recent years the globalization of manufacturing has generated increased competition, forcing manufacturing industries to improve their management systems to sustain a competitive advantage. Industries have now adopted systems that allow for quick changes to product design, procurement, and management of raw materials, manufacturing processes, and human resources.

One such system is lean manufacturing (LM) production. LM is a production philosophy that focuses on the total elimination of waste or activities that do not add value (Womack, 1996). LM produces an operational and cultural environment that is conducive to waste minimization and pollution prevention. It creates an environment of continuous improvement throughout the company and involves all employees. LM is a source of technological capabilities for the creation of competitive advantage for companies, and it focuses on continually improving resource productivity and production efficiency, which frequently translates into less material, less capital, less energy, and less waste per unit of production.

There is evidence that LM produces environmental performance improvements that cost little and result in significant financial advantages to companies. In the early 1980s, Japanese companies began to look at Toyota for its key concepts of LM and the total elimination of waste. The Toyota Production System is the forerunner for LM techniques.

This chapter includes a case study that was conducted in Federal Mogul, an automotive plant specializing in stamping processes for automotive seals. Production lines in the automotive plant were designated as a pilot program for the design of the LM system, which was then to be applied throughout the rest of the company.

In the first phase of the pilot program, a company-wide training program on LM was developed and deployed. The operational staff was trained on LM techniques, importantly on value flow mapping for the detection of waste and Kaizen (Japanese word for "improvement") methods for the development and implementation of continuous improvement. This was done with the intention to involve all staff members in and accelerate the implementation of all the concepts discussed in the Kaizen workshops. During these workshops, employees were encouraged to share their ideas for improvement, which were then implemented with the support of the management team. Kaizen workshops were the key to achieving all of the improvements outlined in this study. The Federal Mogul plant has since adopted Kaizen workshops as a way to achieve world-class manufacturing.

The second phase of the program implemented at Federal Mogul included a system analyses of the production process from the first operation and the flow of information (production scheduling). Operational measures for setup times and activities, work-in-progress, inventory, and production rates were applied. The end research shows that by using the techniques of LM, adequate alternative production systems that generate competitive advantage can be developed.

HISTORY OF LEAN MANUFACTURING

Before World War II, the companies that dominated the market for the automotive industry were General Motors and Ford, creating a monopoly in the automotive market. After World War II, Japan began its operations in Toyota with Taichi Ohno as one of its leaders, who began with the philosophy of continuous improvement (Ohno, 1991).

LM is a philosophy that focuses on the total elimination of waste from suppliers to customers (Womack & Jones, 1996). The LM philosophy seeks only what is necessary when it is necessary. Only the required quantity and products are delivered in the necessary place. LM provides excellent customer value through the creation of processes with zero waste, as detailed by Womack & Jones (1996). LM philosophies are applicable to all phases of production including design, manufacturing, and maintenance.

The phrase "lean production" was coined in the late 1980s by John Krafcik of the international motor vehicle program of the Massachusetts Institute of Technology (MIT) and popularized by James Womack, Daniel Jones, and Daniel Roos in their 1990 book *The*

Machine That Changed the World. The book describes the manufacturing paradigm established by the Toyota Production System. In the 1950s, the Toyota Motor Company pioneered a collection of advanced manufacturing methods that aimed to minimize its resources.

Toyota, inspired by the waste elimination concepts developed by Henry Ford in the early 1900s, created an organizational culture focused on the systematic identification and elimination of all waste from the production process. In the LM context, waste was viewed as any activity that does not lead directly to creating a product or service.

In many industrial processes, such "nonvalue" activity can comprise more than 90% of the total activity as a result of time spent waiting; unnecessary rework of the product; overproduction; wasted movement; and inefficient use of raw materials, energy, and others. Toyota's success from implementing advanced manufacturing methods has led hundreds of other companies across numerous industrial sectors to tailor these advanced production methods to address their operations. Throughout this chapter, the term "lean" is used to describe broadly the implementation of several advanced manufacturing methods. This was the generally accepted name of the system derived from the Toyota Production System. Lean production, or lean manufacturing, began as a means of organizing and managing various operations of production—from suppliers to product delivery—so they require fewer personnel and less time, money, and material to make products with higher quality when the customer demands them.

As manufacturers realized the benefits of lean on the shop floor, the philosophy was incorporated into business processes, indirectly impacting the manufacturing of products, including purchasing, accounting, sales, customer service, distribution, and logistics.

When companies implement several or all of these LM methods, several beneficial outcomes consistently result:

- Reduced inventory levels (raw material, work in progress, finished product) along with associated reduced carrying costs and reduced costs from damage and spoilage of materials
- Decreased material usage (product inputs, including energy, water, metals, chemicals, etc.) by reducing material requirements and creating less material waste during manufacturing
- Optimized equipment (capital equipment utilized for direct production and support purposes) using lower capital and resource-intensive machines to drive down costs
- Reduced need for factory facilities (physical infrastructure primarily in the form of buildings and associated material demands) by driving down the space required for production
- Increased production speed (the time required to process a product from initial raw material to delivery to a consumer) by eliminating unnecessary process steps, movement, wait times, and downtime
- Enhanced production flexibility (the ability to alter or reconfigure products and processes rapidly to adjust to customer needs and changing market circumstances) enabling

the implementation of a pull production and a just-in-time-oriented production system, which lowers inventory and capital requirements
- Reduced complexity (simple products and processes decrease opportunities for variation and error) by reducing the number of parts and material types in products and by eliminating unnecessary process steps and equipment with unneeded features

At the same time, lean implementation consistently fosters changes in organizational culture that exhibit the following characteristics:

- A continual improvement culture focused on identifying and eliminating waste throughout the production process
- Employee involvement in continual improvement and problem solving
- Operations-based focus of activity and involvement among the workers
- A metrics-driven operational setting that emphasizes rapid performance feedback and leading indicators of performance and success
- Supply chain investment to improve enterprise-wide performance
- A whole systems view and thinking for optimizing performance

Lean methods typically target eight types of waste. These wastes are: waiting; inventory; transportation; overproduction; overprocessing; intelligence; motion; and defects, rejects, or scrap.

Waiting. Waiting waste comes from delays associated with stock-outs, processing delays, equipment downtime, and capacity bottlenecks. Whenever an employee is waiting for something, it is costing the company money. Not only is the labor cost per unit higher, but also the machinery is often idle while the wait occurs. If the machinery is not running, it is not being operated at full capacity, reducing overall equipment effectiveness (OEE), and the company is losing efficiency and potential sales dollars.

Inventory. Inventory waste comes from excess raw material, work in progress, or finished goods. Inventory is one of the largest wastes in any industry. It costs money to store the inventory that could be used for necessary items. Materials stored in inventory could quickly become obsolete. Inventory takes up space and potentially causes inefficient operations. Every operation needs some amount of inventory, but by employing LM concepts, the company can reduce inventory to the minimum amount necessary.

Transportation. Transportation of material or product does not add any value to the product. One example of transportation is movement of product through the plant. Mapping the flow of product helps identify some of the transportation waste, as the movement of product often requires transportation.

Overproduction. Overproduction waste is manufacturing items for which there are no orders. Overproduction waste is a very large waste because not only is material lost but also the labor that was put into converting it and the time spent producing it.

Intelligence. An example of the waste of intelligence is not involving everyone in the business for improvement. Operators often have years of experience and know what could be done to improve the business but often are never involved. LM implementation is most successful when it involves everyone in the business improvement, including line operators.

Motion. Examples of motion waste include human motion that is unnecessary. When humans exert strain and transport products long distances in a company, there is wasted motion. The waste of motion occurs in every manufacturing and service industry. An example of this is an employee searching for a tool or a part.

Defects, rejects, or scrap. This waste is the production of off-specification products, components, or services that result in scrap, rework, replacement production, inspection, or defective materials. Any scrap, defect, or reject is considered a waste.

It is interesting to note that the "wastes" typically targeted by environmental management agencies, such as nonproduct output and raw material wastes, are not explicitly included in the list of manufacturing wastes that lean practitioners routinely target for elimination. Some organizations easily make the connection between being environmentally friendly and the lean philosophy and its associated tools. This green frame of mind is ingrained in the culture of the organization that is given the most credit for lean: Toyota. The ideal state of green manufacturing is to use 100% renewable resources with complete use of all materials and no by-products—zero waste. In today's world, much like most value streams, the ideal state is a far cry from reality. In reality, organizations must use a combination of renewable and nonrenewable resources and reusable and nonreusable materials.

Green manufacturing focuses on the following aspects to reduce waste:

- Minimizing the use of nonrenewable resources
- Maximizing the value use of nonrenewable resources when they must be used
- Maximizing the value use of renewable resources
- Maximizing the value use of materials
- Maximizing the reuse of scrap raw materials
- Minimizing total by-products
- Maximizing by-products that can be reused
- Minimizing toxic by-products or by-products that cannot be used
- Minimizing the use of product materials and packaging (including shipping materials)
- Maximizing the reuse of product materials and packaging after use

The term "waste" is defined by James Womack and Daniel T. Jones (1996): "Waste is anything that is not the absolute minimum of time and resources to add value to the product and thus of the highest quality."

They also mention that in a manufacturing process there are activities that add value and cost. It is believed that of all the efforts to transform a product, only 1% adds value to the product, while the other 99% does not add value, but it does add cost to the product. The definition of "value" is that it is only added to the products that the customer pays for, so that activities such as storage, reinventory, handling, and transportation of materials are considered waste. A useful technique for visualizing and analyzing value is the value stream map, which can serve as a starting point to help management, engineers, production planners, and clients see waste and its causes. The value stream map is a simple diagram showing all the steps involving the flow of material and the flow of information needed to manufacture a product from the customer until the order is given. This chart uses a series of symbols to represent each step (Rother & Shook, 1999). Several reports, such as the ones by Spear and Bowen (1999) and O'Corrbui and Corboy (1999), have highlighted the benefits of techniques such as LM, waste elimination, and value stream mapping for continuous improvement.

METHODOLOGY FOR THE CASE STUDY

To demonstrate that the application of continuous improvement technique generates benefits to the company, LM was implemented in a pilot area in an automotive seals manufacturing plant and the results were measured. This study aims to demonstrate that the use of LM creates an adequate alternative production system that generates a competitive advantage by providing a product with value in the eyes of the customer. This case study includes analysis of the production process.

Plant managers, supervisors, engineers, and operators throughout the company cooperated for the case study and provided data. A data management strategy aimed at quality improvement was implemented, based on the Six Sigma methodology of DMAIC (define, measure, analyze, improve, and control) (Michael, 2002). Two additional phases were added the traditional Six Sigma DMAIC methodology: reporting and auditing. Thus, the data management strategy implemented can be represented as seven interconnected phases: define, measure, analyze, improve, control, report, and audit (DMAICRA). These steps can be defined as the following:

- *Define*. Define the problem, define the process of improvement, form a team, and train the team.
- *Measure and analyze*. Collect data to verify the existence of the problem, draw the value stream map, and analyze the data.
- *Improve and control*. Draw the value stream map of the future state, select the area to improve through Kaizen events, select the technical information to be used for improvement, and train for the selected technical information.
- *Report*. Generate a report of findings.
- *Audit and feedback*. Implement audits to ensure the sustainability of the gain and feedback.

Define: Problem Definition in the Case Study

In the area of production, metallic stampings were produced for different product families. The product family that was the subject of the study represents 60% of the total volume of transmission seals. The family for this product consists of four part numbers.

The demand for this product had increased substantially over the past 2 years by approximately 25%. The amount of part numbers also increased about 50%. The demand

Figure 14.1. Seal arrow.

increase had generated the need for having more flexible and agile processes to meet the requirements of customers without increasing the costs.

The problem faced was that the company had not implemented a production system that included all the tools of LM that would lead to continuous improvement.

Overall objective. The main objective of this research was to design and implement the system of LM through the plant and determine how much impact it has on reducing waste.

Specific objectives. The design and implementation of the system of LM has the following specific objectives:

- Define whether the inventory process decreases as a result of implementing LM tools in the current production process.
- Determine the increase in the output rate per hour compared with the current production (in areas of assembly and molding).
- Determine whether the total cycle time can be reduced in the assembly area as a result of applying LM tools to the current production process.
- Determine whether the total shift in the areas of assembly, stamping, and molding are reduced and the extent to which the current process can optimize the shift.
- Determine whether the waste is reduced as a result of implementing LM tools.
- Analyze the differences between the current production conditions (push) and the proposed production manufacturing using lean (pull).

Data Collection and Analysis

In August 2003, the requirements of production were 180,000 stamps per week, with a target productivity of 76.3% in the assembly area. See Figures 14.2 through 14.6 for analyses of the plant's productivity, downtime, waste, and inventory.

As shown by the value stream map in Figure 14.7, the production process has high inventory, which leads to a stagnation of the flow of material, partly due to high shift times. This also causes a decline in the rate of output per hour, a high level of waste and inefficiency,

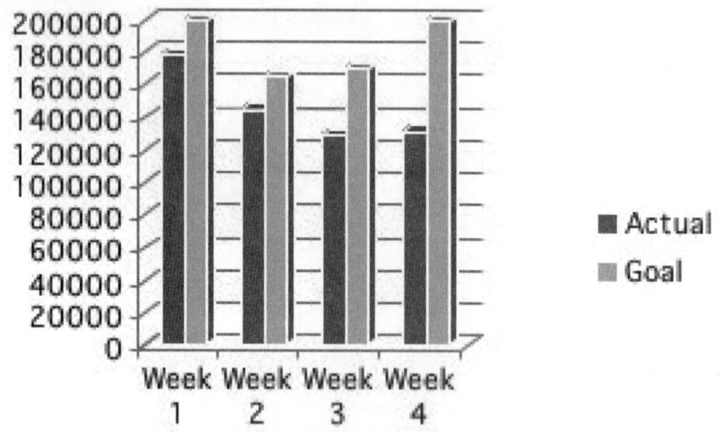

Figure 14.2. Chart productivity in August 2003.

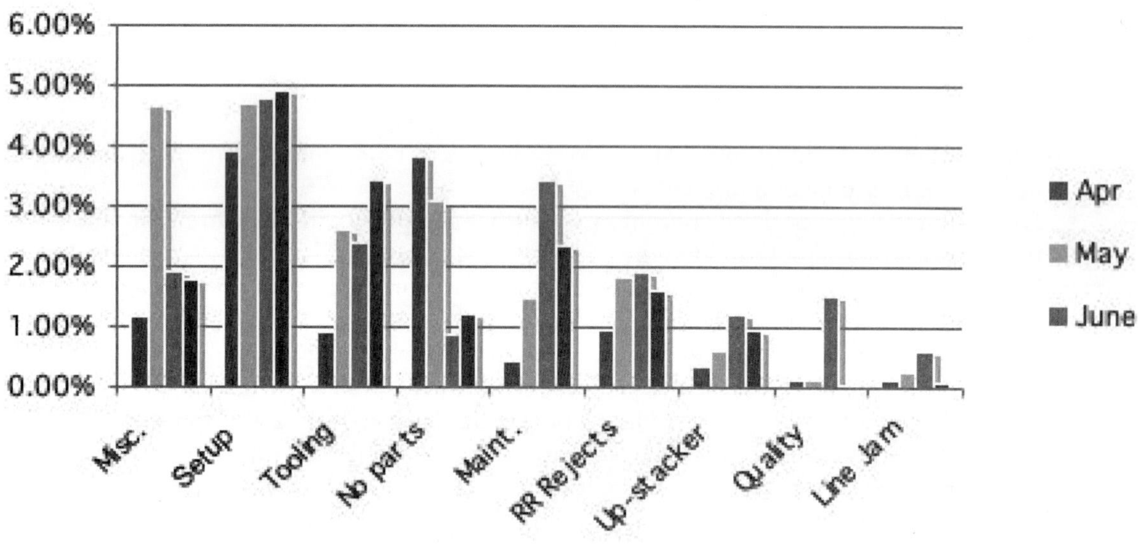

Figure 14.3. Chart downtime assembly area in August 2003.

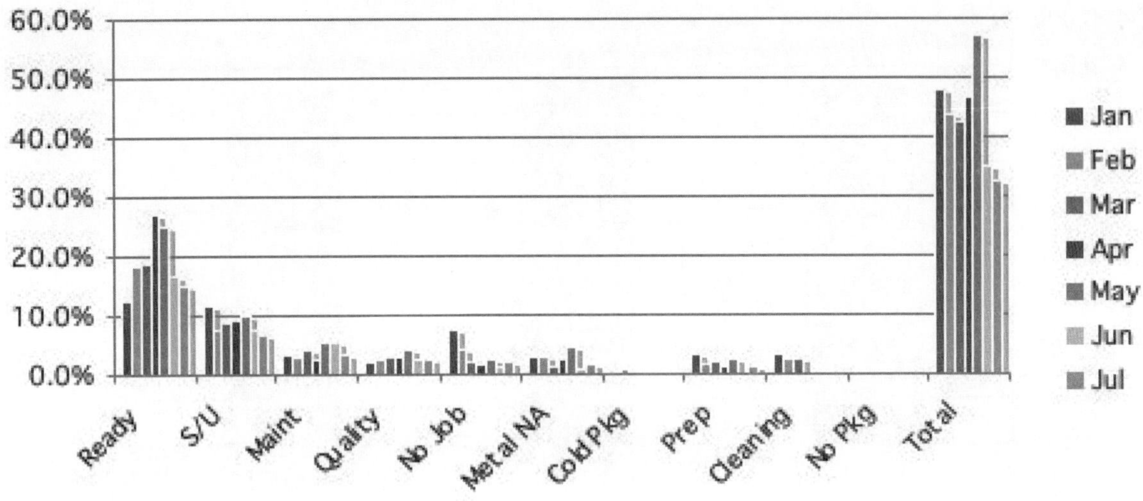

Figure 14.4. Chart downtime molding area in August 2003.

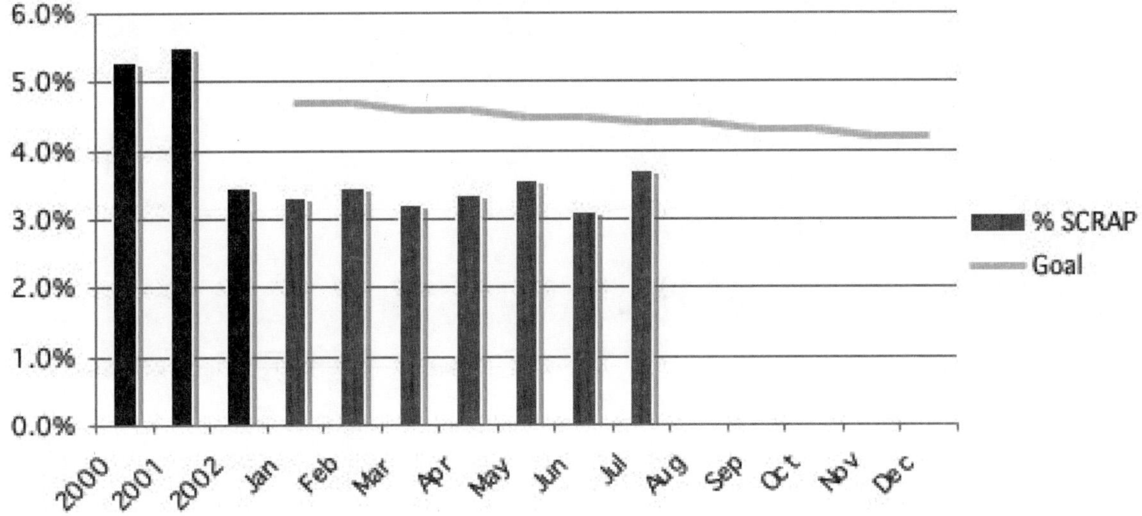

Figure 14.5. Chart of the area of waste molding in January–June 2003.

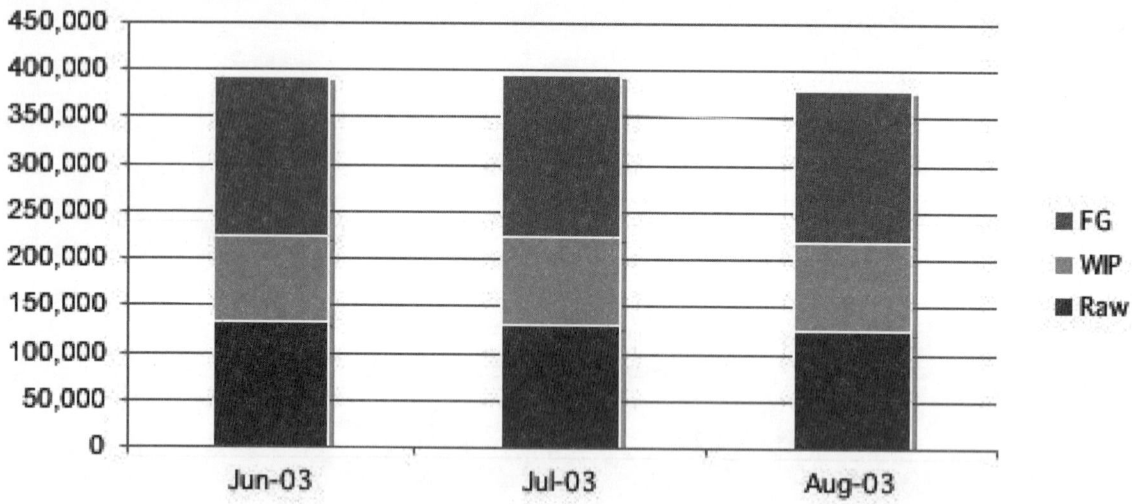

Figure 14.6. Inventory of the product under consideration from June to August.

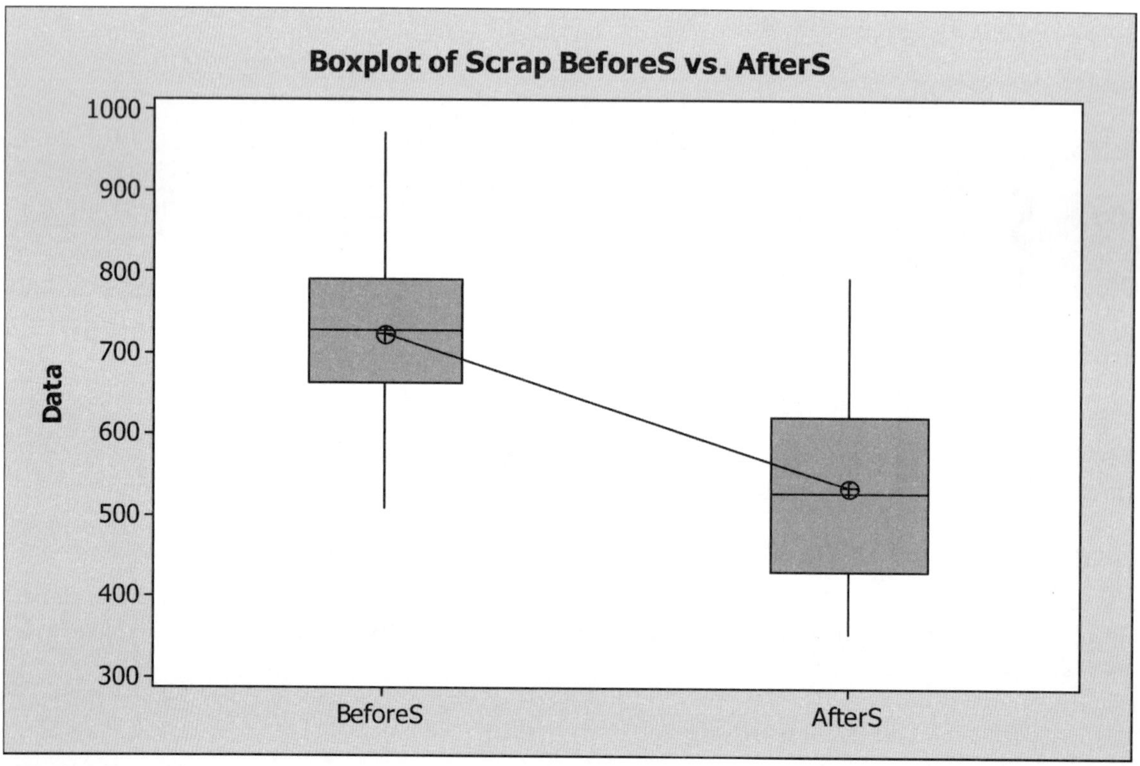

Figure 14.7. Value stream map of the current state (MI).

and a long total cycle time. The requirements for production programming for each of the processes is based only on one report and not the requirements of the next process.

The value stream map shows the following problems: The total cycle time is higher compared to the response time. The total cycle time is 14.3 days, and the response time is 10 days. Accumulation of inventory is high to prevent any adjustments and minimize the differences between the cycle time and the response time. Because of a poor balance in the process, the material is moved in large quantities by forklift. Programming and coordination of production are difficult due to individual workload. Because each machine produces at full capacity and there is a large separation between machines, it is difficult to identify and detect the defects caused by excess inventory. In the areas of assembly and molding, there is wasted time.

Improve and Control

The first improvement was to select and train the staff with the 5S lean manufacturing technique, which is based on the concepts of sorting, straightening, systematic cleaning, standardization, and sustaining discipline. All staff participated in the training. The first three steps of the 5s technique (sorting, straightening, and systematic cleaning) were run on the production floor. These phases eliminated all things on the production floor that were not intended to be in the area, such as folders, boxes, and tables. Separate, defined spaces were created for various materials, and all tools and parts were labeled and identified. Managers and the operators cleaned the work area thoroughly. Standardization, the next step in the 5s process, helped define the limits of the colors and fonts in the automotive seals so that specifications on seals could be followed by all operators consistently. As a final step (sustaining discipline), the 5s technique was reviewed with employees for up to 20 minutes during every shift. This review helped identify conditions in the area such as leaking oil and metal powder, which were reported to maintenance for repair.

The second technical improvement to the production process involved the implementation of the single minute exchange of die (SMED) method. The SMED method provides a rapid and efficient way for employees to set up for and anticipate upcoming processes that follow a given task. Three operators, two engineers, and two maintenance technicians were trained in the SMED method. A video was taken to analyze all the activities involved in production so that the transition from one task to another could be made efficiently and tool and die change activities could be streamlined and coordinated. Preparation of the tooling took around 11 minutes. The operator had to go to the tool crib placed in a shelf area with all the tooling and perform preventive maintenance for the tools, including cleaning the tool. SMED implementation reduced tool preparation time by 5 minutes. Apart from these activities, improved internal adjustments of the die resulted in an average reduction of 12 minutes.

The third technical improvement was a layout design. A "U" shaped layout was designed to improve material flow and eliminate the movement of the operator, as well as to reduce the production area from 618 square feet to 358 square feet. The customer demand for four particular models produced by the plant was 150,000. Given a daily demand of 1428.5 stampings per hour, the Takt time (defined as the ratio of the daily demand and the available time per shift [Rother & Harris, 2001]) was 2.5 seconds per piece. With a production cycle time of 2.1 seconds, utilizing the revised layout, customer demand was easily met.

The fourth technical improvement implemented was the pull system. It was implemented in the area between the molding and painting using containers with a capacity of 2,000 stampings.

The implementation of preventative maintenance, the 5s technique, redesign of the workstations, reductions in operator movements, and improvements in the appearance of the area, helped reduce shift time by an average of 1 hour 45 minutes (105 minutes). Overall, all design changes from implementing LM techniques resulted in a total 30 hours of reduction in time. This reflected in the profit and reduced the amount of overtime required to fulfill production demands.

Report

Table 14.1 shows the productivity and scrap information before and after the implementation of LM.

Audit and Feedback

Audits are implemented to ensure the sustainability of the gains. In order to verify and audit the sustainability of the gain, a hypothesis test was conducted. The following hypothesis for productivity gains was tested.

Null hypothesis: H0: After the implementation of LM, the productivity level was the same as before the implementation.

$$H0: \bar{Y}pb = \bar{Y}pa$$

Alternative hypothesis: Ha: After the implementation of LM, productivity levels increased.

$$Ha: \bar{Y}pb < \bar{Y}pa$$

See Table 14.2. The data indicate that there was an increased productivity level, as the *p* value is equal to 0.000. Figure 14.8 shows the comparisons of the productivity level before and after the lean implementation.

The second hypothesis that was tested was for differences in scrap rate produced from the process before and after implementation of LM.

Table 14.1. Data of Productivity and Scrap, Before and After

	Productivity		*Scrap*	
No.	*Before*	*After*	*Before*	*After*
1	22,350	25,700	827	540
2	17,340	26,800	971	490
3	27,250	28,500	790	627
4	21,200	31,000	742	713
5	24,000	29,500	792	649
6	23,500	28,700	893	390
7	28,250	31,500	650	441
8	18,350	30,800	826	462
9	23,450	28,800	750	374
10	22,000	29,500	748	354
11	13,500	28,000	716	560
12	24,500	31,000	711	589
13	22,300	31,700	691	793
14	21,100	29,800	781	536
15	23,500	32,100	823	674
16	26,000	30,400	676	669
17	16,300	29,700	668	683
18	25,500	28,500	561	399
19	21,800	28,900	654	376
20	24,300	28,000	778	504
21	26,100	30,500	757	519
22	19,000	27,000	798	621
23	21,500	28,100	667	618
24	18,200	31,000	673	403
25	24,850	30,300	770	505
26	1,500	28,000	630	588
27	20,400	29,500	714	590
28	27,300	32,000	519	384
29	24,200	27,500	508	523
30	23,750	29,900	594	508

Table 14.2. Hypothesis Test of Productivity Increase

	N	*Mean*	*StDev*	SE *Mean*
Before *p*	30	21776	5134	937
After *p*	30	29423	1630	298

Difference = *mu* (before *p*) – *mu* (after *p*)

Estimate for difference: –7647.00

95% upper bound for difference: –5984.05

T test of difference = 0 (vs <); *t* value = –7.78; *p* value = 0.000; Degrees of Freedom = 34

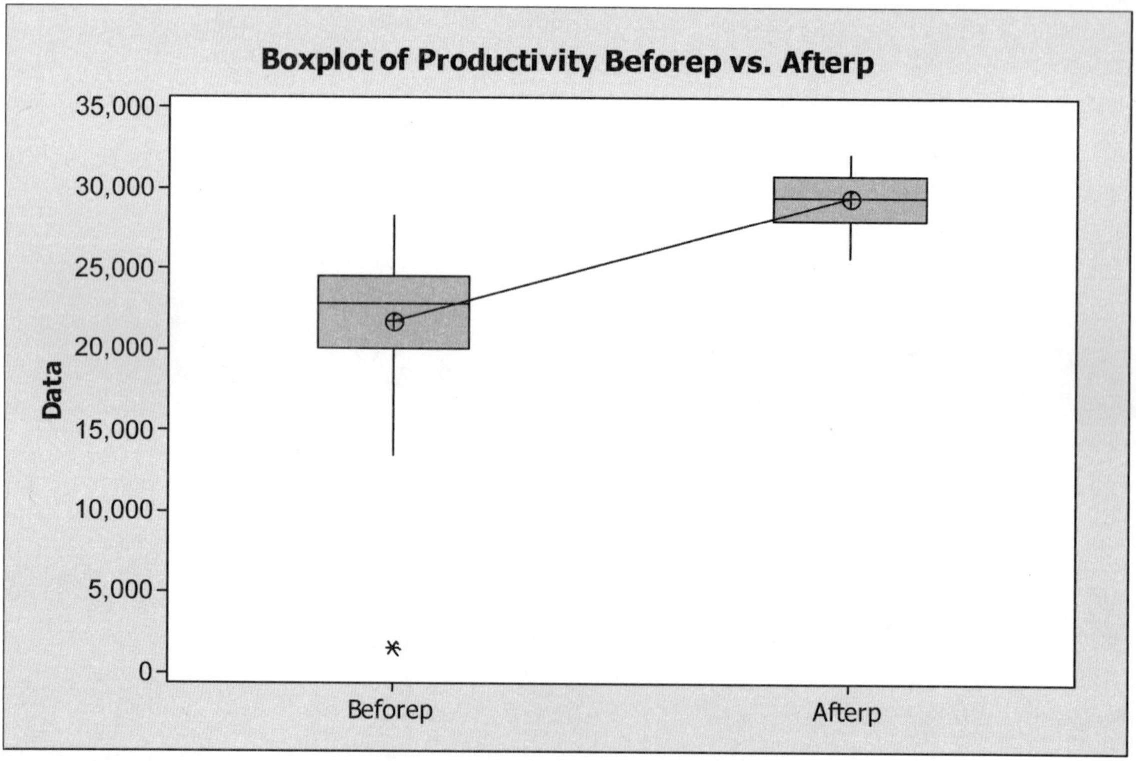

Figure 14.8. Boxplot of productivity.

Null hypothesis: Ho: The level of scrap is the same as before the implementation of LM.

$$Ho: \bar{Y}sb = \bar{Y}sa$$

Alternative hypothesis: Ha: The level of scrap was reduced after the implementation of LM.

$$Ha: \bar{Y}pb > \bar{Y}pa$$

Table 14.3. Hypothesis Test of Scrap

	N	*Mean*	*StDev*	SE *Mean*
Before *S*	30	723	103	19
After *S*	30	536	115	21
Difference = *mu* (before *S*) – *mu* (after *S*)				
Estimate for difference: 186.533				
95% lower bound for difference: 139.267				
T test of difference = 0 (vs >): *t* value = 6.60; *p* value = 0.000; DF = 57				

The data indicate that there was reduced scrap level, as the p value is equal to 0.000. Figure 14.9 shows the comparisons of the scrap level before and after the lean implementation.

According to data from Table 14.4, one can conclude that productivity increased from 79.4% to 96.5%, scrap was reduced from 3.38% to 1.42%, inventory was reduced from \$375.00 to \$274.00, and the total cycle time was reduced from 14.3 days to 8.2 days.

CONCLUSIONS AND RECOMMENDATIONS

The implementation of LM in the Federal Mogul plant in Summerton, South Carolina, was a success. This was largely due to the support of management and the great support of all the people who participated in the workshops and the Kaizen engineers who provided leadership and ideas for implementing this philosophy.

As mentioned previously, the principles for the implementation of LM (as described by James P. Womack and Daniel T. Jones [1996]) are as follows:

- Define product value in the eyes of the customer.
- Identify the value stream and eliminate waste.
- Improve flow of material.

- Utilize a pull production system.
- Continuously improve the processes and products.

Implementation of LM principles helps to add value to the production process and identify areas where tools could be utilized more efficiently.

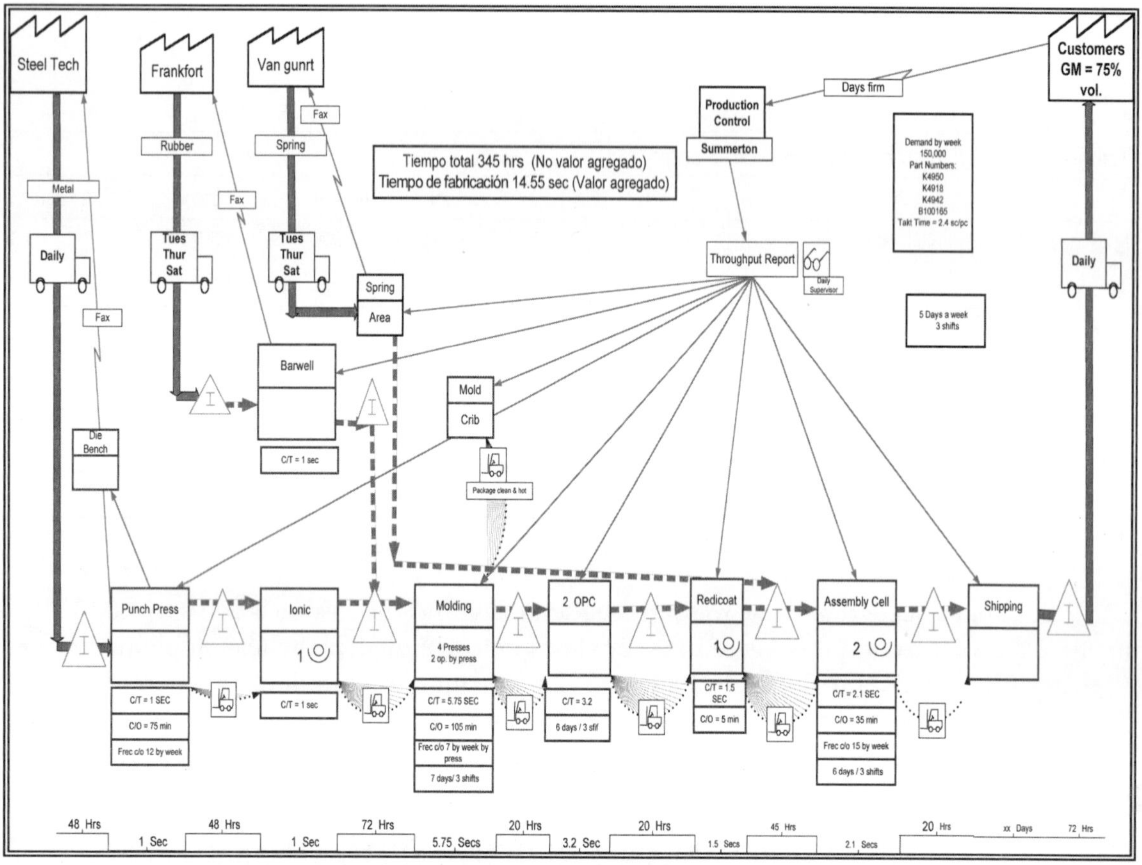

Figure 14.9. Boxplot of scrap.

Table 14.4. Data Before and After Implementation

Before	*After*
Productivity 79.4%	Productivity 96.5%
Waste 3.38%	Waste 1.42%
Inventory $375,000	Inventory $274,000
Total cycle time 345 hours (14.3 days)	Total cycle time 197 hours (8.2 days)

Conclusions

The organizational structure required for implementing lean manufacturing is important to define and understand. The roles and responsibilities of people required for implementing lean manufacturing are the following:

- *Steering committee*. Group executives are responsible for addressing the program, performance evaluation, and resolution of critical issues and approval of new projects.
- *Leader in lean manufacturing*. The leader is responsible for the efficient operation of the processes of project management, monitoring of project progress on the plan, and creation of consolidated reports.
- *Sponsor*. The sponsor is responsible for the progress of projects in his or her area to the management committee, ensuring appropriate levels of resources for projects, and supporting team leaders in resolving critical issues.
- *Team leader*. Team leaders are responsible for coordinating various projects, ensuring compliance dates, communicating with sponsors and with the leader, and developing new ideas for improvements (project charter).
- *Leading project manager*. The leading project manager is responsible for developing detailed work plans for project management, successfully implementing the project, and escalating critical issues that require support.
- *Member*. A member is responsible for compliance shares allotted by the project leader, management, and the balancing of daily activities.

Recommendations

To implement lean manufacturing in an organization, the following steps are recommended:

- Conduct a regular review of the behavior of the data from the key metrics such as productivity, scrap, and inventory in order to detect deviations from specifications.
- Fill all templates and data collection tools for data such as downtime, waste, productivity, and so on.
- Continue with the routines of order and cleanliness, the five Ss, in all areas to help create a better working environment.
- Establish a regular program and training for operational staff, supervisors, manufacturing engineers, and maintenance staff to promote the interest and motivation for continuous improvement.
- Continue to search for the best production practices to eliminate as much waste as possible and to maintain a world-class plant and to make your plant a candidate for an expansion and not a failure.
- Revise the annual model changeover routines for updating and training the staff.
- Develop new leaders who are well trained in the philosophy and tools of lean manufacturing.

REFERENCES

Bednarek, M. (1996). *The key to the design and implementation of manufacturing cells*. Mexico City, Mexico: Technology Group.

Brassard, M., & Ritter, D. (1994). *The memory jogger: A pocket guide of tools for continuous improvement and effective planning goals*. Salem, NH: QPC.

Feld, W. M. (2001). *Lean manufacturing tools, techniques, and how to use them*. Boca Raton, FL: St. Lucie Press.

George, M. L. (2002). *Learn six sigma: Combining six sigma quality with lean production speed*. New York, NY: McGraw-Hill.

Goldratt, E. M. (1990). *What is this thing called the theory of constraints*? Great Barrington, MA: North River Press.

Goldratt, E. M., & Cox, J. (1992). *The goal: The process of ongoing improvement*. Great Barrington, MA: North River Press.

Harris, R., Harris, C., & Wilson, E. (2003). *Making materials flow*. Cambridge, MA: Lean Enterprise Institute.

Jones, D., & Womack, J. P. (2002). *Seeing the whole: Mapping the extended value stream*. Cambridge, MA: Lean Enterprise Institute.

Liker, J. K. (2004). *The Toyota way: 14 management principles from the world's greatest manufacturer*. New York, NY: McGraw-Hill.

Masaaki, I. (1992). *Kaizen: The key to Japan's competitive success*. New York, NY: McGraw-Hill.

O'Corrbui, D., & Corboy, M. (1999). The seven deadly sins of strategy. *Management Accounting, 10*, 1–5.

Pande, P. S., Neuman, R. P., & Cavanagh, R. R. (2002). *The six sigma way team fieldbook: An implementation guide for process improvement teams*. New York, NY: McGraw-Hill.

Rath & Strong. (2000). *Six sigma pocket guide*. Lexington, MA: Rath and Strong Management Consulting.

Richard, L. (2002). *The lean enterprise memory jogger goal*. Salem, NH: QPC.

Rother, M., & Harris, R. (2001). *Creating continuous flow: An action guide for managers*. Cambridge, MA: Lean Enterprise Institute.

Rother, M., & Shook, J. (1999). *Learning to see: Value stream mapping to create value and eliminate muda*. Cambridge, MA: Lean Enterprise Institute Inc.

Sekine, K. (1992). *One piece flow: Cell design for transforming the production process*. Portland, Oregon: Productivity Press.

Smalley, A. (2004). *Creating level pull: A lean production-system improvement guide for production-control, operations, and engineering professionals*. Cambridge, MA: Lean Enterprise Institute.

Spear, S., & Bowen, K. (1999). Decoding the DNA of the Toyota production system. *Harvard Business Review*, 77(5), 95–106.

Spiegel, M. R. (1991). *Statistics*. New York, NY: McGraw-Hill.

Taiichi, O. (1995). *Toyota production system: Beyond large scale production*. Barcelona, Spain: Productivity Press.

Womack, J. P., & Jones, D. T. (1996). *Lean thinking*. New York, NY: Simon & Schuster.

INDEX